SpringerBriefs in Biochemistry and Molecular Biology

For further volumes:
http://www.springer.com/series/10196

Rodrigo J. Carbajo · José L. Neira

NMR for Chemists and Biologists

 Springer

Rodrigo J. Carbajo
Structural Biochemistry Laboratory
Centro de Investigación Príncipe Felipe
Valencia
Spain

José L. Neira
Inst. de Biología Molecular y Celular
Universidad Miguel Hernández
Elche
Spain

ISSN 2211-9353
ISBN 978-94-007-6975-5
DOI 10.1007/978-94-007-6976-2
Springer Dordrecht Heidelberg New York London

ISSN 2211-9361 (electronic)
ISBN 978-94-007-6976-2 (eBook)

Library of Congress Control Number: 2013939831

Springer is part of Springer Science+Business Media (www.springer.com)

Preface

This book in the Springer Brief Series intends to be an easy and concise introduction to the field of nuclear magnetic resonance or NMR, which has revolutionized life sciences in the past 20 years. A significant part of the progress observed in scientific areas like Chemistry, Biology or Medicine can be ascribed to the development experienced by NMR in recent times. Many of the books currently available on NMR deal with the theoretical basis and some of its main applications, but they generally demand a strong background in Physics and Mathematics for a full understanding. This book is aimed to a wide scientific audience, trying to introduce NMR by making all possible effort to remove, without losing any formality and rigor, most of the theoretical jargon that is present in other NMR books. Furthermore, we have provided illustrations showing all the basic concepts using a naive vector formalism, or using a simplified approach to the particular NMR-technique described. Our intention has been to show simply the foundations and main concepts of NMR, rather than seeking thorough mathematical expressions.

The book is organized into four chapters. Chapter 1 introduces the basis of NMR; furthermore, a concise description of how the magnet works and the instrumental set-up is provided at the end of this chapter. Chapter 2 describes the different spectroscopic parameters in NMR such as the chemical shift and the spin coupling, which will be used in the following chapters. Chapter 3 describes some basic experiments in NMR, starting from the basic one-dimensional experiment and continuing with the principles of multi-dimensional experiments. The last chapter describes the most successful applications of NMR in Biochemistry, from structural elucidation to drug-discovery and metabolomics.

We believe that the target audience for this book will be undergraduate and graduate students willing to give their first steps in the NMR field. The book can be used as a complement for courses of Analytical Techniques or Advanced Techniques in the Grades of Biochemistry, Biotechnology, Pharmacy or Biology. It has been also written for lecturers in the above disciplines who want to take a first look on general concepts of NMR or get more information about the subject for their respective research fields. We hope that even readers not familiar with the topic, but curious enough to increase their general knowledge, will find this book useful.

Trying to keep in mind the ample width of the intended audience, the book has been revised by scientists with different backgrounds and expertises. We are indebted to Francisco N. Barrera (Yale University), Ana Cámara (Almería University), and Tennie Videler (Cambridge University) for their many helpful suggestions and corrections. We are also deeply indebted to Thijs van Vlijmen and Sara Germans, our editors in Springer, for their patience, help, and corrections to the early version of the manuscript. However, any error left in the text the reader is now holding, is our sole responsibility.

We are delighted to receive any suggestions from the readers to correct, modify, or improve the clarity of the text. Moreover, we hope that you all enjoy the book and will use it as a gateway for understanding the fascinating and burgeoning world of NMR.

Valencia, Spain, March 2013   Rodrigo J. Carbajo
Elche, Spain   José L. Neira

Contents

Abbreviations

aa	Amino acids
acac	Acetylacetonate
ADC	Analogue-to-digital converter
AF	Audio-frequency
B_0	External static magnetic field
CD	Circular dichroism
COSY	Correlation spectroscopy
CP	Cross-polarization
CPD	Composite pulse decoupling
CPMG	Carr Purcell Meiboom Gill
CSA	Chemical shift anisotropy
CW	Continuous wave
DEPT	Distortionless enhancement by polarization transfer
DMSO	Dimethyl sulfoxide
DQF	Double-quantum filter
DSS	4,4-dimethyl-4-silapentane-1-sulfonic acid
FID	Free induction decay
FT	Fourier transformation
HMBC	Heteronuclear multiple bond correlation
HMQC	Heteronuclear multiple quantum coherence
HOHAHA	Homonuclear Hartman-Hahn
HSQC	Heteronuclear single quantum coherence
HTS	High throuput screening
INADEQUATE	Incredible natural abundance double-quantum transfer experiment
INEPT	Insensitive nuclei enhanced by polarization transfer
IR	Infra-red
J	Coupling constant
M	Magnetization
MAS	Magic angle spinning
MQC	Multiple quantum coherence
NMR	Nuclear magnetic resonance
NOE	Nuclear Overhauser effect

NOESY	Nuclear Overhauser effect spectroscopy
PCA	Principal component analysis
PFG	Pulse field gradient
RF	Radio-frequency
rmsd	Root mean square deviation
ROE	Rotating frame effect
ROESY	Rotating frame Overhauser effect spectroscopy
S/N	Signal-to-noise ratio
SAR	Structure-activity relationships
SPT	Selective population transfer
SQC	Single quantum coherence
ssNMR	Solid-state NMR
STD	Saturation transfer difference
SW	Spectral width
TMS	Tetramethylsilane
TOCSY	Total correlation spectroscopy
trNOE	Transfer NOE
TSP	Trimethylsilyl propionate
UV	Ultraviolet
Water-LOGSY	Water-ligand observation with gradient spectroscopy
τ_c	Correlation time

Chapter 1
The Basis of Nuclear Magnetic Resonance Spectroscopy

Abstract Nuclear magnetic resonance (NMR) has transformed the research areas of Chemistry, Biochemistry and Medicine, but much of its fundamentals remain obscure for the non-initiated. NMR is a technique based on the absorption of radiofrequency radiation by atomic nuclei in the presence of an external magnetic field. In this chapter, we describe the physical basis of phenomena within NMR spectroscopy from both a quantum-theory perspective and a classical view, to provide any prospective user the basic concepts underlying the technique. The spectroscopic notion of energy level population is described, as well as a basic introduction to the theory and mechanisms of spin relaxation. The application of radiofrequency pulses to produce the NMR signal, its conversion to the frequency domain by Fourier Transform and the typical instrumental set-up of magnetic resonance are also covered.

Keywords Dipolar interaction · Gyromagnetic · Magnet · Nuclear magnetic resonance (NMR) · Nuclear spin · Quantum levels · Radiofrequency pulse · Relaxation · Spectroscopy

1.1 Introduction

The aim of this chapter is to give a brief summary of the physical basis of nuclear magnetic resonance (NMR) spectroscopy. The theoretical description provided here is not exhaustive, and the interested reader is encouraged to refer to more advanced texts for further information (Ernst et al. 1987; Günther 1995; Wüthrich 1986; Derome 1987; Claridge 1999; Cavanagh et al. 1996; Keeler 2006; Evans 1996; Sanders and Hunter 1992). In our treatment of NMR theory, we shall introduce the classical vector approach and a naïve quantum mechanical approach, trying to avoid, for the sake of simplicity, the complicated mathematics behind the theory.

R. J. Carbajo and J. L. Neira, *NMR for Chemists and Biologists*,
SpringerBriefs in Biochemistry and Molecular Biology,
DOI: 10.1007/978-94-007-6976-2_1, © The Author(s) 2013

NMR is based on the magnetic properties of atomic nuclei, which may be considered to be composed of spinning particles in the simplest model. NMR uses an effect which is well-known in classical physics: when two pendulums are joined by a flexible axle and one of them is forced into oscillation, the other is forced into movement by the common flexible support and the energy will flow between the two. This flow of energy is most efficient when the frequencies of the two movements are identical: the so-called *resonance* condition. Another example of the resonance condition is found in radio antennas. A radio antenna responds to broadcast radiofrequency signal through the movement of electrons, which shift up and down in the antenna at the same frequency as that of the broadcast signal. In both examples, resonance (the same frequency) is the key, as it is in NMR. In NMR, instead of electrons or pendulums, there are nuclear spins; therefore, we need first to understand how electromagnetic radiation interacts with those spins.

1.2 Physical Principles of NMR Spectroscopy

1.2.1 The Basis of NMR Spectroscopy: A Vector Approach

For an introduction to the classical formalism in NMR, the interested reader can have a look in the literature (Farrar and Becker 1971). From a classical physical perspective, a charge travelling circularly around an axis builds up a magnetic moment (dipolar moment or magnetic dipole), μ, with a direction perpendicular to the plane defined by the circular movement of the charged particle. The faster the charge travels, the more intense the induced magnetic field and the stronger the magnetic dipole. Atomic nuclei contain positively charged particles which have such rotating movement and consequently generate nuclear magnetic dipoles. In the presence of an external static magnetic field, B_0, the magnetic moment μ of the nuclear particles will orient either with (parallel) or against (anti-parallel) B_0. This interaction causes a movement of precession of the nuclear dipoles around the B_0 axis (Fig. 1.1a), in a manner analogous to a gyroscope in a gravitational field; this precession is a consequence of the rotation about the own nuclear axis of the particle. The angular frequency of this precession, ω, depends on: (i) the strength of B_0; and (ii) the properties of each particular nucleus. This frequency is called the *Larmor frequency* and is given by $\omega = 2\pi\upsilon = \gamma B_0$, where the proportionality constant γ is called the *gyromagnetic* or *magnetogyric* ratio and υ is the *frequency*. It is important to emphasise that the precession is only possible in the presence of an external B_0.

The Boltzmann distribution dictates that the most stable and lowest energy state in a system will be the most populated at equilibrium. In this case, the nuclear magnetic dipoles oriented with B_0 have the lowest energy. However, the difference in energy with the less stable state (dipoles oriented against B_0) is small, leading to a small difference in population. Although minor, the population imbalance leads to a net nuclear magnetic moment, which is the sum of the dipole moments

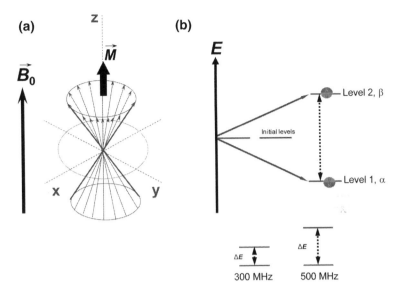

Fig. 1.1 The physical basis of NMR. **a** Classical view of NMR. Spins precess around the external magnetic field B_0; the net magnetization of the system (the sum of all the nuclear magnetic dipoles of the sample) is shown along the z-axis as M. The *thin arrows* are the magnetic dipoles of each nucleus. **b** Quantum view of NMR. The two quantum levels originated as a consequence of B_0 (*left* side) are shown (for the sake of simplicity, it has been assumed that the nuclear spin has only two possible orientations, as for the ^1H atom); the orientation of the two spins is indicated as small circles in each level showing the polarity of the nuclear magnets. The *dotted arrow* indicates the jump from one quantum level to the other. The arrow on the *left* side of the figure represents the energy. To allow for a comparison, at the *bottom* of the figure two split levels are shown for a 300 and 500 MHz magnets (the figure is not to scale)

(vectors) of all nuclei present in the sample. This net nuclear magnetic moment is called *magnetization* (M), and it is represented by a vector parallel to the static magnetic field B_0 (Fig. 1.1a). The length of the M vector is proportional to the population difference between the parallel and antiparallel dipolar moments. If an appropriate amount of energy is applied, M will move from its equilibrium position to become oriented with B_0 ($-M$). The corresponding absorption of energy experienced by the nuclei in that action is measured by NMR and the energy is in the order of that of radio waves.

1.2.2 The Basis of NMR Spectroscopy: A Naïve Quantum Approach

From the perspective of quantum mechanics, the magnetic resonance phenomenon takes place due to an intrinsic property of nuclei: the *spin* (the rotation around its own nuclear axis). To introduce spin we shall consider a simple particle like the

Table 1.1 Predicting the nuclear spin (*I*)

Number of protons (Z)	Number of neutrons	Nuclear spin (*I*)
Even[a]	Even	0
Even	Odd	1/2, 3/2,...
Odd	Even	1/2, 3/2,...
Odd	Odd	1, 2,...

[a] 0 is considered here an even number

familiar electron. The electron has spin, which is the source of its angular momentum (due to its rotation movement), taking a value of ½. Since in quantum mechanics each moment *l* has $2l + 1$ associated values, the value of ½ means that there are $(2 \times ½) + 1$ different possible levels, with quantum numbers ½ and $-½$. In the absence of an external magnetic field, both levels have the same energy (they are degenerate). In the presence of an external magnetic field, the two levels become separated by a difference in energy ΔE.

Similarly to the electron, atomic nuclei have *spin angular momentum*, represented by *I*. The value of *I* for each nucleus depends on its atomic structure (number of protons and neutrons) (Table 1.1). As a consequence of the spin, the nucleus has an associated *dipole moment* (μ), which takes the value $\mu = \gamma I$. Analogous to the electron spin angular momentum, the nuclear spin angular momentum is quantized and can only take $(2I + 1)$ discrete values, given by the quantum number m_I. These discrete values are energetically degenerate (as with the electron levels) in the absence of any external magnetic field, but in the presence of B_0, quantized (discrete) energy will separate the two levels. For instance, a proton (^{1}H) has $I = ½$ (like the electron) and its spin can adopt either of two orientations (½ or $-½$), which are degenerate in the absence of an external field (B_0). The state with $m_I = ½$ (↑), known as α, is oriented parallel to the external B_0, whereas that with $m_I = -½$ (↓) and denoted by β, represents the antiparallel orientation to the external B_0 (Fig. 1.1b). For those nuclei with spin ½ and $\gamma > 0$, the α state has a lower energy than the β state. On the other hand, nuclei with $I = 1$, like ^{14}N, will show three energy levels, in the presence of B_0, corresponding with m_I taking the values 1, 0, -1. Nuclei with $I = 0$ (such as ^{12}C) do not have dipole moment μ, no energy levels to be split and therefore no NMR signal.

The energy difference between the levels is determined by $\Delta E = \mu B_0$, and presents values within the radiofrequency spectrum. Therefore, the larger B_0, the larger the energetic separation and the resulting population differences (Fig. 1.1b, bottom). With currently available magnets the energy gap is quite small (at room temperature, for an 11.1 T magnet the difference between the populations for protons is in the order of 1 in 10^{-5}) which is the basis for the inherent low sensitivity of NMR spectroscopy. To induce a transition between the energy levels, it is necessary to apply an electromagnetic radiation with an amount of energy given by the Planck equation. Absorption of applied radiation, that satisfies the resonance condition (of the Larmor frequency), causes a spin-½ nucleus to flip from the α (low energy) to the β (high energy) state. These changes in energy levels are referred as *NMR transitions*. A net absorption of energy is possible as long as the population of nuclear spins in the higher energy level is smaller than in the lower one. When, in the presence of an applied B_0, the populations of the two levels become equal, the system is saturated and no further absorption of energy will ensue. We shall discuss later (Sect. 1.3) how a system can return to the equilibrium state, where the lower energy level is more populated.

1.2.3 The Nuclei in NMR

If the number of both the protons and neutrons in a given nuclide are even, such as in the most abundant isotope of carbon $_6^{12}C$, then $I = 0$ and thus, there can be no NMR signal (Table 1.1). On the other hand, those nuclei with an odd number of either protons or neutrons will have a half-integer spin (i.e. $\frac{1}{2}$) and those presenting an odd number for both protons and neutrons will have a spin $> \frac{1}{2}$ (e.g. 1, 3/2, 5/2, etc.). The number of neutrons varies within each chemical element giving rise to its isotopes, thus for each individual element there might be one or more NMR-active isotopes. Among the $\frac{1}{2}$ spin nuclei, the most interesting nuclides are ^1H and ^{13}C because of the ubiquity of these elements in the structure of natural molecules. The ^{15}N, ^{19}F or ^{31}P (Table 1.2) nuclei are also much used and chemically interesting although any isotope with $I \neq 0$ can in principle be detected by NMR. A comprehensive description of the NMR properties of all the nuclei in the periodic table can be found in the literature (Harris and Mann 1978; Mason 1987). It is important to note that the value of γ is different among the nuclei. Since the separation among the levels in the presence of an external magnetic field (Fig. 1.1b) relies on γ, it is clear that the larger a nuclide's γ value, the larger the separation between levels, and therefore the better its sensitivity in NMR experiments.

Table 1.2 Magnetic properties of selected nuclei

Isotope	I	$\gamma \ 10^7 \ (T^{-1} \ s^{-1})$[a]	Abundance (%)
^1H	1/2	26.75	99.985
^2H	1	4.11 ($2.8 \ 10^{-3}$)	0.015
^{14}N	1	1.93 ($1.0 \ 10^{-3}$)	99.63
^{15}N	1/2	-2.71	0.37
^{12}C	0		98.89
^{13}C	1/2	6.73	1.108
^{16}O	0		99.96
^{17}O	5/2	-3.627 ($-2.6 \ 10^{-2}$)	0.037
^{19}F	1/2	25.18	100.0
^{31}P	1/2	10.83	100.0
^{33}S	3/2	2.05 ($-5.5 \ 10^{-2}$)	0.76

[a] Quadrupolar moment indicated in $10^{-28} \ m^2$ (for nuclei with $I > 1/2$)

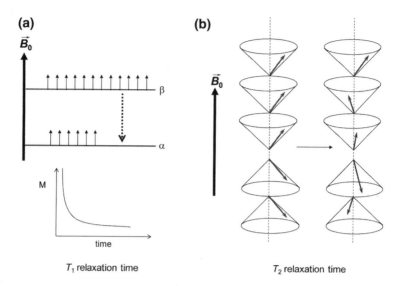

Fig. 1.2 Spin relaxation. **a** The longitudinal spin–lattice relaxation time (T_1) is the time constant for the exponential return of the magnetization to the equilibrium population. The arrows represent the population of spins. The graphics at the *bottom* represent the exponential behaviour that the recovery of the equilibrium population (represented by M) follows (indicated by the *thick dotted arrow* in the *upper* panel). **b** The transversal spin–spin relaxation time (T_2) is the time constant for the exponential return of the spins to a random distribution of their phases (the angle formed between the precessing spins and the external field B_0). No change in population is observed in this type of relaxation, and therefore no energy is transmitted from the spins to their neighbourhood

Nuclei with $I > \frac{1}{2}$ (Table 1.2) display two characteristic features. First, they will show ($2I + 1$) quantum states which lead to more complicated transitions than the single jump between the two levels of a spin-$\frac{1}{2}$ nucleus. Second, they show an asymmetric distribution of charge leading to a nuclear electric quadrupole, which gives them the common name of *quadrupolar* nuclei. An electric quadrupole can be visualized as a charge distribution similar to the four lobes of an atomic *d*-orbital (in this analogy, an electric dipole would correspond to a *p*-orbital), which is a non-spherical charge distribution. Such arrangement results in a strong interaction with the electric field gradients generated by the non-uniform distribution of local electronic charges at the nucleus. Moreover, the quadrupole is coupled to the nuclear μ that is also present in these nuclei. All these effects lead to quadrupolar relaxation effects that are responsible for the broadening and poor resolution of their NMR signals (Sect. 1.3). This broadening, together with their often reduced sensitivity (small γ) or low natural abundance, are the main reasons for the relatively low popularity of the NMR investigation on quadrupolar nuclei.

1.3 Spin Relaxation

In this section, we describe how the populations of spins recover after the excitation and return to their equilibrium distribution. Importantly, these so-called *relaxation mechanisms* can also provide information on the dynamics of the molecules. For clarity purposes, we shall restrict the discussion to a spin-½ nucleus.

In the equilibrium state, the system only possesses longitudinal magnetization (along the z-axis: the M vector in Fig. 1.1a), and not a transverse one (that is, in the xy-plane). From a quantum perspective, apart from the uneven populations in the two quantum levels, this means that the individual spins, at equilibrium, have a random distribution of phases of their wave-functions (the states α or β of the particular spin). After the absorption of energy two steps are needed for the return to equilibrium: the restoration of the initial populations (that is, M aligned with the z-axis) and the randomization of the phases of the wave-functions of the nuclear spins. In NMR, the absorption of energy is produced by an externally oscillating electromagnetic field (that is, a radiofrequency; Sect. 1.4), and relaxation is also induced by oscillating magnetic fields within the molecules (for example, other spins in the same or different molecules within the sample). Hence, relaxation in NMR occurs by magnetic interactions. This variation in the internal oscillating magnetic fields is modulated by molecular motion and it is a property of the local environment of each particular spin.

At this stage, it is important to recognize the difference between the relaxation processes in NMR and other spectroscopies, where the recuperation of the equilibrium population occurs by collisions with other molecules and dissipation of energy by heat release. The collision is extremely slow and inefficient in NMR, as collisions involve mainly the electronic shells of the nuclei. This leads to NMR excited states with lifetimes in the range of milliseconds to seconds compared to excited molecular states with lifetimes as short as μs to ns in other spectroscopic techniques. This extended lifetime of excited states is crucial to the success of NMR, since it does not only mean that the signals are narrower than in other spectroscopies (according to the Heisenberg uncertainty principle), but also allows the manipulation of the spins after the excitation, leading to the myriad of different experiments available which result in the richness of the technique and its applications.

1.3.1 Spin–Lattice Relaxation Time

We have already described how for our spin-½ nucleus in the presence of an external field, B_0, two quantum levels α and β will separate. Electromagnetic radiation with the adequate υ will cause transitions to occur between those two levels, with spins jumping from the lower to the higher one. If the resonant absorption continues, it seems evident that after a certain time the number of spins at the β level will match that in the α state. When the two populations become equal, the intensity of the NMR signal decreases because a "saturated" state has been reached. The situation is similar to that experienced by youngsters (the spins), trying to enter into a popular disco (the upper level) where a delicious drink/beat/music… (the energy) is provided; if the disco is full (saturation) no net entry of people (the intensity) within the venue can take place.

Non-radiate processes (those which do not cause emission of electromagnetic radiation, but rather emission of heat) by which β nuclear spins can become α spins again occur. Interactions between the spins and their surroundings, also known as the *lattice*, lead to loss of energy. These non-radiate processes, by which the population of nuclear spins returns to equilibrium (in the classical model, magnetization recovering its *longitudinal* position parallel to B_0), follow an exponential behaviour, dependent on the *spin–lattice* relaxation time (or longitudinal time), represented by T_1 (Fig. 1.2a). It is important to emphasise that T_1 describes the time taken by the magnetization to recover the position aligned with the z-axis, and during that time the population of the system reverts to the uneven distribution of the equilibrium populations. Finally, it is important to bear in mind that some authors describe the spin–lattice relaxation rate ($R_1 = 1/T_1$) instead of a spin–lattice relaxation time T_1 (Sanders and Hunter 1992).

1.3.2 Spin–Spin Relaxation Time

We shall continue to consider a spin-½ system, this time with three spins precessing aligned along B_0 and two precessing against B_0 (note that we are not saying that this situation corresponds to the equilibrium population) (Fig. 1.2b, left). If each spin has a somewhat different Larmor frequency (due, for instance, to a dissimilar chemical environment or to field inhomogeneity), they will gradually fan out and, at thermal equilibrium, the entire set of vectors will lie at random angles to the direction of the B_0 (Fig. 1.2b, right). It is important to point out that the population of the two spin-states has not been altered, and that there is no transfer of energy to the surroundings in this case, as described in spin–lattice relaxation. The time governing this exponential behaviour, which returns the system to a random arrangement of phases, is called the *spin–spin (transversal)* relaxation time T_2 (the second factor mentioned above when we were describing how to restore the equilibrium population). In some texts, the spin–spin time is again represented as the relaxation rate $R_2 = 1/T_2$ (Sanders and Hunter 1992). We shall return to the meaning of *transverse* for T_2 later when we describe the effect of pulses on magnetization (Sect. 1.4).

Taken together, we can say T_1 acts as the enthalpic component to restore the equilibrium population (since there is a rearrangement of the populations), and T_2 works as the entropic component in the restoration of the equilibrium (since it involves disorder of the phases of the spins).

1.3.3 Sources of Variation in Local Fields

Variation in the local magnetic field is one of the factors responsible for the relaxation mechanisms, but what causes these? In this section we shall describe the sources of the local fields, whose variations cause both spin–lattice and spin–spin relaxation processes. The key point in the discussion is that local fields fluctuate both in magnitude and direction, depending on the orientation and movement of the molecule with respect to the applied B_0.

1.3.3.1 Dipole–Dipole Interactions

The mechanism of nuclear spin relaxation lies predominantly in magnetic interactions, of which the most important one is *dipole–dipole* relaxation or *dipolar interaction*, in short. This type of interaction is also the source of the *nuclear Overhauser enhancement* or NOE effect (Chap. 2). Dipolar interactions can be easily grasped using the classical view of nuclei as bar magnets. The dipolar coupling between two magnets depends on the distance between them, as well as on the external B_0 (Fig. 1.3a). If we replace these two magnets by the spins of two nuclei, A and B, within a molecule that tumbles randomly in solution, this random movement will alter the internuclear vector r_{AB}. Additionally, the local field experienced by any nucleus will fluctuate randomly as the result of variations in its surroundings. Random motion of a molecule occurs due to molecular collisions, and can be characterized by the *correlation time*, τ_c, which is usually defined as the average time it takes a molecule to rotate one radian ($\sim 57°$). Long τ_c times correspond to slow tumbling of the nuclei (as found for macromolecules), and short τ_cs correspond to fast tumbling (as displayed by small molecules).

   Which kind of local magnetic fields can induce a transition from β to α (that is, spin–lattice relaxation)? The ideal field would be one that fluctuates at a frequency close to the Larmor frequency of the involved nuclei. Such field can arise from the tumbling motion of the molecule. If the frequency of this molecular tumbling is slow compared to the Larmor frequency, it will cause an oscillating magnetic field that fluctuates too slowly to induce transitions and consequently T_1 will be long. Conversely, if the frequency of the molecular tumbling is faster than the Larmor frequency, this fluctuating field will oscillate too fast to induce transitions, and therefore T_1 will be long again. Only when the frequency of the tumbling rate is about the same as the Larmor frequency, T_1 will be short, and the molecule will show fast relaxation. It is important to bear in mind that the rate of molecular tumbling increases with temperature and with the decrease of the viscosity of the solvent (Fig. 1.3b).

   What are the most effective local dipolar interactions to induce spin–spin relaxation? The most favourable ones would be those presenting an interaction at a relatively slow variation rate. Under the influence of this slowly-changing local magnetic field, each nuclear spin in the sample remains in its local magnetic

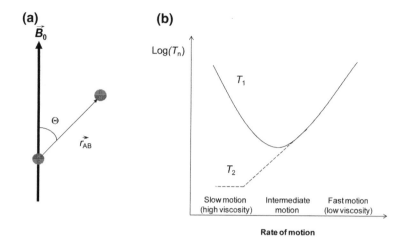

Fig. 1.3 Variations in the local fields. **a** Physical parameters on which the dipolar–dipolar interaction relies: the distance between spins r_{AB}, and the angle Θ formed between the r_{AB} vector and the external magnetic field B_0. **b** Variation of the longitudinal and transversal relaxation times with the molecular movement rate (either by tumbling or migration through the solution), which depends, in turn, on changes of temperature or viscosity

environment, giving the spins enough time to randomize their orientations. Obviously, this will preferentially occur at low temperatures, in high viscosity conditions, or in very rigid molecules (e.g. solid-state samples). Conversely, if the molecule moves very fast (for instance, at high temperature or low viscosity) T_2 increases and the randomization does not take place. That is, slow molecular motion corresponds to short T_2 times, and fast molecular motion corresponds to long T_2s. T_2 is always smaller than T_1, but at high temperatures both relaxation times become close to equal (Fig. 1.3b).

For nuclei relaxing mainly through dipolar interactions, the line-width of the signals in the spectrum is determined by: (i) the proximity of other nuclei; and (ii) the value of τ_c. For instance, the line-width of the signals in the NMR spectra of large biomolecules is very broad due to their associated slow overall tumbling. The relaxation via dipolar interactions can also arise from the magnetic interaction of a nuclear spin with an unpaired electron, which also behaves as a miniature magnet. This effect is known as *paramagnetic relaxation*. As the magnetic moment of an electron is over 1000 times that of a proton, this relaxation mechanism is very efficient; in fact, this dipolar interaction is used to deliberately reduce the relaxation times of spins in molecules with a high number of isolated nuclei by adding agents containing paramagnetic species, such as derivative of metals of the series d and f of the periodic table (e.g. Cr(III) in Cr(acac)$_3$).

1.3.3.2 Chemical Shift Anisotropy

The local variation of fields is a direct result of the anisotropic electronic environment of molecules, which "shields" or "unshields" nuclei from the external magnetic field in an orientation dependent manner. This varying field can stimulate relaxation if it is sufficiently strong. Chemical Shift Anisotropy (CSA) is important, for instance, in atoms having sp^2 hybridization and therefore non-spherical electronic charge distribution. The CSA contribution to spin relaxation correlates quadratically with B_0 (that is, CSA becomes more significant at higher fields), but is usually less relevant than dipolar interactions. In biomolecules, ^{13}C, ^{15}N and ^{31}P nuclei have significant CSA contributions.

1.3.3.3 Quadrupolar Relaxation

In low-symmetry environments quadrupolar nuclei ($I > \frac{1}{2}$; Sect. 1.2.3) experience non-zero electric field gradients, which depend on the molecular orientation with respect to B_0. This anisotropic interaction is modulated as the molecule tumbles at an appropriate frequency (as in the dipolar interactions), inducing flipping of nuclear spin states, and therefore, spin relaxation. Relaxation of quadrupolar nuclei is governed by the features of the quadrupole nucleus itself and the electric field gradient, which depends on the local environment. Therefore this relaxation mechanism is electric-dependent rather than magnetic. In general, the larger the quadrupole and the electric field gradient, the more efficient the relaxation mechanism is, due to shorter T_2s. On the other hand, quadrupolar nuclei in highly symmetrical environments (tetrahedral or octahedral structures) have electric field gradients close to zero, and the quadrupolar relaxation mechanism is ineffectual. For instance, the NMR spectrum of ^{14}N ($I = 1$) in tetrahedral NH_4^+ has a line-width in the range of 1 Hz, but pyramidal NH_3 (a less symmetrical environment than a tetrahedron) has line-widths close to 100 Hz.

1.3.3.4 Spin-Rotation Relaxation

Small molecules, or groups of atoms in a large molecule (e.g. methyl groups), can rotate rapidly in low-viscosity solvents, thus creating a molecular magnetic moment due to the rotating electronic and nuclear charges. The local field associated with this magnetic moment is modulated by molecular collisions and movements, providing another nuclear spin relaxation mechanism: the so-called *spin-rotation* relaxation. This mechanism is more effective in small, symmetrical molecules or freely rotating groups, and the relaxation efficiency increases with the reduction of rotational correlation times.

1.3.3.5 Scalar Relaxation

Relaxation may also be induced when a nucleus is scalarly coupled (Chap. 2) to
another one which fluctuates rapidly. This fluctuation might originate from:
(i) chemical exchange of the coupled nucleus; or (ii) fast T_1 of the fluctuating
nucleus (common in quadrupolar nuclei). A practical example is observed for
nitrile groups ($-C\equiv N$), where the carbon is covalently bound to the quadrupolar
^{14}N nucleus resulting in a significant broadening of the ^{13}C signal.

1.4 Pulse Techniques

In most spectroscopic techniques such as UV or IR, experimentalists obtain spectra
by varying the wavelength of the excitation source. A similar approach can be
taken in NMR: varying the wavelength of the radiofrequency source will
sequentially induce transitions of different atomic nuclei between their quantum
levels (Fig. 1.1b). Alternatively, the same effect can be accomplished by keeping
the excitation radiofrequency constant while varying the external field. For technical reasons, the latter approach is more practical in NMR and it is applied
extensively.

In the early times of NMR (in the fifties to seventies of the last century), NMR
magnets applied the slow-passage *continuous-wave* (CW) method, which involved
the variation of the external B_0 to reach the resonance condition (the name comes
from the continuous application of one radiofrequency while resonance is
observed). Modern pulse methods excite the entire radiofrequency region of the
nucleus of interest by applying a single variation of radiofrequency power (a pulse)
and the resulting signals are recorded after the radiofrequency is turned off.
A Fourier transformation (FT) is required to convert the resulting signal into the
familiar NMR spectrum and, therefore, pulse-NMR is often called FT-NMR. One
of the best analogies for comparing the old and new ways to acquire NMR spectra
is the detection of the spectrum of sounds (vibrations) of a bell-ladder (a set of
nearby bells with different sizes, joined by their closed termini). If we sequentially
strike each of the bells with a unique hammer, we will collect the single sound
(frequency) of a particular bell. Conversely, if we strike the whole set of bells
forming the ladder, we shall produce a clang contributed by all the frequencies of
each particular bell. Our challenging mission will be to disentangle the particular
frequencies within that clang. In NMR, the "bells" are our nuclear spins, and the
"clang" of frequencies is the magnetization decay of the spins as they return to
equilibrium after excitation with a radiofrequency pulse (the "hammer").

1.4.1 How a Pulse Works: The Time and Frequency Domains

In this section, we describe first how the pulse affects the magnetization (M) of a system in the presence of B_0, and how we obtain the time and frequency domains of an NMR spectrum.

1.4.1.1 The Effect of a Radiofrequency Pulse: Changing the Observation System

Since most of the tiny magnetic dipoles within the atoms–but not all, due to the Boltzmann distribution at equilibrium–are aligned with the external field, there is a net magnetization M along the direction defined by B_0 (in the classical view). If we apply a short (in the order of microseconds) radiofrequency perturbation (a *pulse* or the so-called B_1) along the x-axis (Fig. 1.4a), with a frequency matching the separation energy between the quantum levels (or, in other words, the Larmor frequency in the classical view), the spins aligned initially with the z-axis will move away from that axis and onto the xy-plane.

The B_0 is larger than the B_1 field by many orders of magnitude. For instance, in the case of the ^{1}H nucleus the B_0 field corresponds to precession frequencies of hundreds of MHz, whereas the B_1 is about 1–20 kHz. The B_1 field is not static, but it rather rotates about the z-axis (the axis of the B_0) with a frequency that closely matches the Larmor frequency of the spins that originate from M. Under the influence of these two magnetic fields, M will start precessing around B_1 (with a frequency $\omega = \gamma B_1$) simultaneously as it does around B_0 (with $\omega_0 = \gamma B_0$), in a motion called *nutation*. In order to pave the way for a better understanding of the effect of B_1, we need to change our system of reference. Before, we have been observing the precession of the spins around B_0 from our "laboratory system" (the so-called "frame system"; that is, the system from where we have applied the B_0). If we change our observation system to one that rotates at the Larmor frequency, ω_0, we shall not notice the rotation of the spins, in the same way we do not feel the rotation of the Earth around its axis, because we are within the "Earth-observation system". In this *rotating frame of reference*, those spins whose Larmor frequency is exactly ω_0 will experience a null effective field, only feeling the effect of B_1 while precessing about its axis as long as B_1 is turned on. Spins with a slightly different precession frequency than the Larmor one will experience a precession around an effective magnetic field resulting from the sum of the "felt" B_0 and the applied B_1 (Fig. 1.4b).

Since B_1 is only applied for a short period of time, the effect on M is only a partial revolution around the B_1 axis, with the result of M being "tipped" towards the y-axis. The tip angle α (in radians) is given by $\alpha = \gamma B_1 t_p$, where t_p is the duration of the radiofrequency pulse (this expression is easily obtained from the equation of a uniform rotational movement: $\omega = \alpha/t_p$, and then $\alpha = \omega\, t_p = \gamma B_1 t_p$).

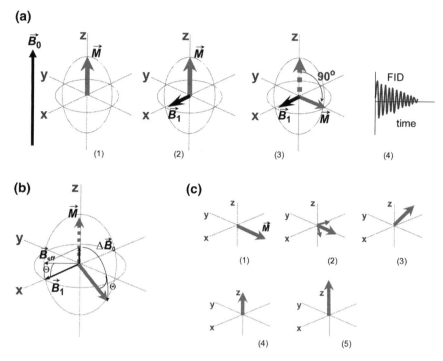

Fig. 1.4 Pulses in NMR. **a** How does a pulse work? Starting from the system at equilibrium (1) in the presence of an external magnetic field B_0 (left-hand side), a $(90)_x$-pulse (shown as B_1) is applied along the x-axis (2), with the effect of tilting the magnetization M (grey arrow) away from the z-axis and onto the xy-plane (3). This tilted magnetization evolves with time to recover its equilibrium position, and this decay renders the FID (4). **b** Scheme of the resulting field "felt" by spins whose precession frequency does not match the Larmor frequency of the "rotating frame" (the so-called *off-resonance* spins). In these spins, a component of B_0 (ΔB_0, the off-resonance value) adds to the pulse (B_1); the resulting "effective" field (the addition of ΔB_0 and B_1) is B_{eff}, which forms an angle Θ with the x-axis; this B_{eff} leaves the spins out of the y-axis, and therefore out of the xy-plane after a $(90)_x$-pulse has been applied. **c** Behaviour of M (*grey arrow*) after the application of a $(90)_x$-pulse within the "rotating frame of reference": (1) immediately after the $(90)_x$-pulse; (2) field inhomogeneities appear (for reader's sake only two isochromatic spins are shown here) giving rise to T_2-effects; (3) T_1 relaxation tilts M back to the z-axis; (4) due to T_2-effects, when M returns to the equilibrium position along the z-axis, the phases of the spins are different (that is, a complete loss of magnetization in the xy-plane has occurred) and M has not recovered the full original length at equilibrium; (5) complete restoration of the equilibrium M occurs after sufficient time is allowed

The pulse is described according to: (i) the tilting it produces in M; and (ii) the axis on which it is applied. For instance, the term $(\pi/2)_x$ or 90_x indicates a pulse directed along the x-axis, which results in a 90° tilting (or $\pi/2$ radians) on the y-axis. The time length of a 90_x pulse for ^{1}H in a 500 MHz spectrometer is around 5–10 μs, with a field B_1 of 10^{-4} T, compared to the B_1 field of 10^{-8} T used in CW spectrometers. It is important to note that when the pulse is applied all spins experience the same effect, generating what some authors call *coherences* (in the quantum view, the phases of all the wave-functions of the excited spins are the same) (Günther 1995; Wüthrich 1986; Derome 1987; Claridge 1999; Cavanagh et al. 1996; Keeler 2006; Evans 1996) (Sect. 1.3).

We can further elaborate on the meaning of t_p in the above expression by using the Heisenberg uncertainty principle (energy-time). This principle states that there is a minimum uncertainty (given by the Planck constant) in the simultaneous specification of both the energy of a system (and hence, the range of frequencies excited in the NMR) and the duration of the measurement (that is, t_p). If t_p is very short, then the range of frequencies excited is very large (i.e. there is a large uncertainty in the range of frequencies excited); this is commonly called a "hard" pulse. On the other hand, if t_p is very long (a so-called "soft" pulse), then the range of frequencies excited is very small (its uncertainty is very small) and we have a small range of nuclei (large selectivity) being excited. This soft pulse can be used to selectively excite some frequencies in the spectrum while leaving the rest unaffected. A common application consists in the selective saturation of the water signal in biomolecular samples, where the soft pulse is applied in a continuous way.

1.4.1.2 The Time Domain

For nuclei whose Larmor frequencies match exactly the frequency of the rotating frame of reference, once the perturbation on the spins in the form of the pulse is removed, the tilted M will stay forever on the y-axis in the absence of relaxation effects. However, with time the different spins forming the net M will start to disperse with their different precession frequencies, returning to the z-axis due to T_1-relaxation effects. On top of this, the inhomogeneities in B_0 cause different regions of the sample to "feel" slightly different B_0 fields, and therefore different Larmor frequencies (which is related to T_2). Thus, the projection of M on the y-axis will shrink according to T_1- and T_2-relaxation times (Sect. 1.3) leading to an exponentially decreasing signal termed the *free induction decay* (or in short, FID). The FID is named for the fact that the collected signal is "free" of the influence of B_1, it is "induced" in the coil and "decays" back to equilibrium. When M is on the y-axis and the magnetization starts to relax its way back to the z-axis due to T_1, the projection of M on the xy-plane will start decreasing as well as fanning out in the plane. The rates of both processes (decrease and fanning) depend on T_2 and hence the name of *transverse* as opposed to *longitudinal* relaxation, which occurs along the z-axis. In modern NMR instruments the setup of the receiver coil is usually along the x or y-axis (Sect. 1.5.4), and the FID is therefore directly related to T_2. The mathematical expression $\exp(-t/T_2)$ defines the envelope of the FID giving the maximum value possible for the transversal magnetization: the exponential decrease of the y-axis magnetization component of M will induce the change in the voltage of the receiver coil, allowing the recording of the FID. However, it is observed in all cases that the FID decays much more rapidly than the value obtained from that envelope. This faster decay is due to real magnet inhomogeneity, which is, in turn, the principal source of T_2-relaxation effects.

Since (i) the effects of both relaxation times (decreasing the projection of the M vector along the x and y-axis (T_2-relaxation) and increasing the projection of the magnetization vector along the z-axis (T_1-relaxation)) add together; and (ii) T_2-relaxation time is shorter than T_1 (and therefore, the y-axis component of M decays to zero before the Boltzmann distribution is recovered) we shall have in the end an intermediate position (Fig. 1.4c) where M is on the z-axis, but its full equilibrium value is not completely recovered. However, after some extra time M will fully return to its equilibrium value with its vector length completely recovered.

It is important to remark that each point in the FID contains information about every resonance that will be observed in the final spectrum (see below). The more data we collect and the longer the period we observe the signal, the higher the resulting spectral resolution. However, if for any reason the signal decays too fast (because of a short T_2) we will not be able to record more useful information in the FID regardless of how long we keep collecting data. The FID is equivalent in our bell analogy above to the "clang" emitted by the bell-ladder, but instead of having all the frequency information as a sound, it is time-encoded, the so-called *time-domain*. Traditionally, the FID is recorded for a specific t_2 time (note that this is an "observation" time, different to the relaxation time T_2) (Fig. 1.4a, right).

1.4.1.3 The Frequency Domain: The Fourier Transformation

The time domain signal is not interpretable as it does not have the usual features of a spectrum of signal intensity *versus* frequencies or wavelengths. The NMR frequency domain (or as we shall see, the NMR spectrum with chemical shift information, Chap. 2) can be obtained using a Fourier transformation (FT). In this, the individual contributions of the different spins to the FID can be separated by means of applying FT to the time-dependent signal (which relies on the mathematical conclusion that any time-dependent function is equivalent to a frequency-dependent function) (Günther 1995; Wüthrich 1986; Derome 1987; Claridge 1999; Cavanagh et al. 1996; Keeler 2006; Evans 1996; Sanders and Hunter 1992). The mathematics and computation of the FT have been described in (Bracewell 1978). The FT will allow the splitting of the "clang" into the individual frequencies of each "bell" within the bell-ladder. In practice, the FT of the time-domain is achieved by using the Cooley-Tukey algorithm, which is reasonably fast even in computers with a moderate amount of available memory. The result of the FT is a complex function of frequencies in the sense that every point in the final spectrum has a real and an imaginary part. When this spectrum is stored onto a computer, the set of coefficients does not contain the final spectrum in the so-called *absorption mode* (all the peaks showing a maximum) and further processing is necessary (Chap. 3).

Several FIDs from simple spin systems and their corresponding transformed spectra are shown in Fig. 1.5. To illustrate, we consider the proton signal of the H_2O resonating at exactly the Larmor frequency (that is, the frequency of B_1 is

(a) On-resonance

(b) Off-resonance

(c) More than two spins

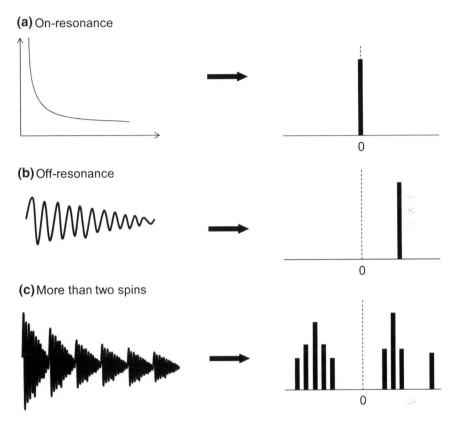

Fig. 1.5 FIDs and their spectra. **a** A simple FID (*left*) of a single on-resonance nucleus and the corresponding Fourier Transform (FT) spectrum (*right*, in bold). **b** FID and FT spectrum of a single spin system whose resonance frequency is not the same as that of the pulse B_1 (off-resonance nucleus). **c** FID of a system composed of several spins with different resonance frequencies and the schematic resulting FT spectrum

on-resonance). The spectrometer frequency is subtracted electronically from the observed signal prior to digitization, and as the detector is placed in the rotating frame, we obtain a simple exponential decay, where the time constant of the exponential is T_2 (Fig. 1.5a). In this case, the spectrum renders a single resonance at zero (0) Hz frequency. In a second situation, where the Larmor frequency of the nucleus is different to that of B_1 (that is, the frequency of B_1 is *off-resonance*), the FID has sinusoidal oscillations of constant frequency whose envelope is a decreasing exponential with time constant T_2. The resonance will appear slightly shifted from the spectrometer frequency (above or below 0 depending on the difference between the spectrometer frequency and the Larmor one of the nucleus) (Fig. 1.5b). If we have two or more spins with the same or different resonance frequencies, we shall have a signal which is the sum of the different components (Fig. 1.5c), and the difference between the frequency signals is the same as the difference in absolute frequencies. Therefore, the above discussion on the aspect and the shape of the spectra suggests that the relative frequencies of each nucleus are detected, although the absolute frequencies of the observed nuclei are the same. This apparently arbitrary choice of the reference frequency does not cause any problem, as NMR frequencies are conveniently calibrated (Chap. 2). It is important to indicate that, prior to the FT, the acquired data can be manipulated to enhance its appearance or information content. We shall describe these "cosmetic" methods in Chap. 3.

1.4.2 Multipulse Experiments: Measurement of the T_1- and T_2-Relaxation Times as Examples

The above discussion is a simplification: modern NMR is not a single-pulse technique. The richness and power of modern NMR relies on the application of series of pulses between interleaved delays during which the magnetization of nuclear spins evolve. These pulses are applied in combinations with different lengths and phases (that is, along different axis), and to different nuclei at the same time or between the time delays. A series of pulses applied to extract required information from a molecule is called a *pulse sequence*. The idea of applying different pulses with different phase angles is of central importance in NMR; combining the collected data from sets of offsets of experiments in an appropriate manner allows removal of imperfections and selection of relevant signals in a process called *phase cycling*.

Here, we shall describe briefly two examples of pulse sequences, applied to a single type of nucleus. In Chaps. 3 and 4, we shall describe pulse sequences combining the magnetization of several nuclei. The sequence examples described here have been designed for the measurement of T_1 and T_2-relaxation times:

(a) *The inversion-recovery experiment*
    The experiment consists on the application of two pulses: the first of 180° and the second, after a certain delay τ, of 90° applied both along the x-axis. Thus, the pulse sequence can be written as: $(180)_x$-time-$(90)_x$-acquisition (Fig. 1.6a). Initially, the magnetization M is aligned along the z-axis and the $(180)_x$ pulse inverts it into the –z-axis. With the application of the following $(90)_x$, we tilt the whole magnetization back on the xy-plane, to be precise on the y-axis. By varying the time between both pulses, we can study how long it takes for the magnetization along the z-axis to recover, and therefore measure T_1.

(b) *Spin-echo experiment*
    The simplest sequence is $(90)_x$-time-$(180)_x$-time-acquisition, although several modifications have been described (Derome 1987; Claridge 1999) (Fig. 1.6b). After the initial $(90)_x$ that tilts M on the y-axis, the magnetization starts fanning out due to the different chemical environments of each nuclei and the intrinsic inhomogeneities of B_0. The application of a perfect $(180)_x$ takes the

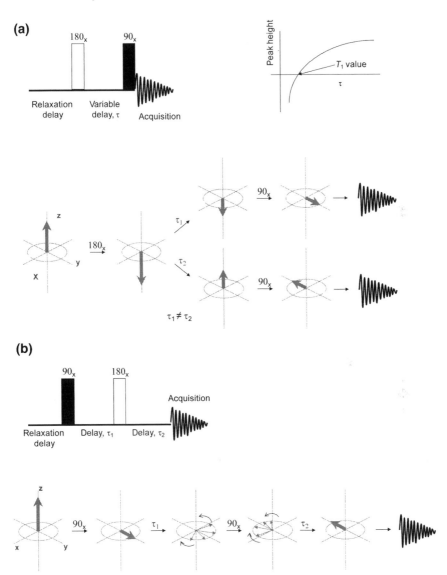

Fig. 1.6 Multi-pulse experiments to measure relaxation times. **a** Inversion-recovery pulse sequence represented with the vector model. *Top left* is the pulse sequence of the experiments (the 90° pulse is shown as filled rectangles and the 180° pulse is shown as *blank rectangles*). *Top right* is the graphical representation of peak intensity *versus* the variable delay τ, which yields T_1 at the intersection point with the x-axis. **b** The spin-echo sequence (Carr-Purcell). On the top, the pulse sequence is shown. On the *bottom*, the graphical representation of the behaviour of the spins according to the vector model is shown. The *arrows* indicate spins which are split due to the T_2-effects; the curved arrows around those spins indicate their circular movement

magnetization vectors on the other side of the xy-plane, and after a time-delay of the same duration as the one between pulses, the vectors will be aligned again along the $-y$-axis, where they will be detected. Varying the time (delay) between pulses will provide a measurement of T_2.

1.5 Practical Aspects of NMR

In this section, we describe some of the practical and instrumental aspects of high-resolution, solution-state NMR spectroscopy. We do not cover aspects such as performance-checking of the spectrometer, determination of its sensitivity or optimization of the signals line-shape, since those experimental aspects are already described in the manufacturer's instructions. Any reader interested in gaining a deeper knowledge on NMR practical aspects is referred to specialized books (Derome 1987; Claridge 1999; Berger and Braun 2003).

1.5.1 The Magnet

Figure. 1.7 shows a schematic overview of a modern NMR spectrometer where the static magnetic field is provided by superconducting materials. The sample is introduced within the magnet as indicated. We shall describe first how the magnet works and in Sect. 1.5.2 how the frequencies from the probe are handled.

The main solenoid producing the B_0 field is placed in a liquid helium bath (at a temperature of 4 K); under these conditions, the electrical resistance of the coil is essentially null. The helium container is surrounded by a liquid nitrogen Dewar

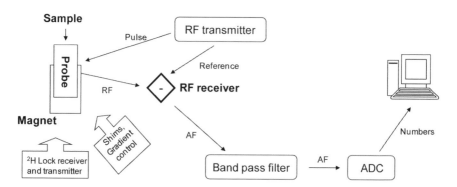

Fig. 1.7 Components of an NMR instrument. Schematic representation of a modern NMR instrument. The receiver reference frequency (RF) is subtracted to leave only audio frequencies (AF) which are digitised in the ADC

vessel (77 K) designed to reduce the loss of He (Fig. 1.8a). This second vessel is isolated from the exterior by a high vacuum chamber. Once the system is energized, the magnet operates indefinitely, free from any external power source. The basic maintenance requires periodical refilling of nitrogen (every 7–10 days) and of liquid helium (every 4–10 months depending on the magnet and the manufacturer). Additional cryo-coils are placed in the He-bath to partially correct any field inhomogeneity.

The magnet is vertically intersected by the room-temperature shim tube. This contraption houses a collection of electrical coils (controlled by software) known as *shim coils* (Sect. 1.5.4) that generate their own small magnetic fields (Fig. 1.8). The shims are used to trim the B_0 and to eliminate residual inhomogeneities of the field. The adjustment of these electrical coils is specific to each sample. The actual NMR sample is contained within a cylindrical glass tube inserted into a plastic or ceramic spinner which is introduced into the magnet from the top of the shim tube using a flow of compressed air or N_2 that drops the tube and spinner into the probe-head (Fig. 1.8b). The sample is in solution and its volume and final concentration depends on the particular application (Günther 1995; Derome 1987; Claridge 1999). For some NMR experiments and low fields (200–300 MHz), the sample is spun at 10–20 Hz to average to zero the field inhomogeneities in the xy-plane, improving signal resolution. However, the spinning introduces additional signal modulations known as *spinning sidebands*, and it is not normally applied at higher fields. At the exact centre of the magnetic field sits the head of the probe, the heart of the NMR spectrometer where the cylindrical tube containing the sample is placed.

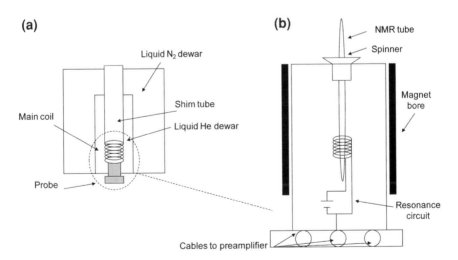

Fig. 1.8 Components of the magnet. **a** Schematic representation of the magnet, showing the He and N_2 chambers, the main coils and the probe-head. **b** Schematic view of the bore with the sample tube, within the spinner, placed in the probe-head. The cables of the preamplifier are also routed to the wobble-unit

1.5.2 The Probe

The NMR probe has two key functions. First, it is the means by which the radiofrequency power is transmitted to the sample for the perturbation of the spin system populations (that is, it creates the pulse). And second, it is required to detect the electromotive force generated by the rotating bulk magnetization, which is in the order of 10^{-9}–10^{-5} V (at the lower end, these values are similar to the amplitude of the electronic noise introduced by the coil). Therefore, the probe houses both the transmitter and receiver coils of the electromagnetic radiation (in modern NMR probes the same coil acts as transmitter and receiver) and associated circuitry, such as the optional gradient coils (Sect. 1.5.6). Probes can be manufactured in various sizes depending on the diameter of the sample tube that it fits around, with the most common being 5 mm, although other sizes are available (e.g. 3, 8 or 10 mm).

Most probes contain two coils: an inner coil tuned to deuterium (the *lock*, Sect. 1.5.3) and a second frequency, whereas the outer coil is tuned to a third and sometimes a fourth frequency. The deuterium channel is placed on the inner coil on most probes as the proximity to the sample increases the sensitivity and renders shorter pulse-lengths. For probes designed to detect ^{1}H (the so-called *inverse probes*) the proton is the second frequency on which the inner coil is tuned. For probes not designed for ^{1}H detection, the second nucleus the inner coil is tuned to can vary.

The receiver coil is at the centre of the magnetic field and the sample in the NMR tube is completely surrounded by this coil. The resonance circuitry inside the probe also contains capacitors that require tuning every time the sample is changed. The probe is connected to a preamplifier, which performs an initial amplification of the NMR signal. The preamplifier contains a wobbling unit required for tuning and matching the probe for each sample. In the wobbling unit the frequency is continuously swept back and forth through the resonance of the nucleus to be observed, covering a bandwidth of 4–20 Hz. A large influence on the tuning and matching of the probe is the dielectric constant of each individual sample, which affects the dielectric constant in the inner coil region and therefore its impedance; high ionic strength (high salt concentration) will significantly increase the length of the pulses.

The amplifiers are low-noise audio-amplifiers which boost the amplitude of the incoming signal to yield a larger undistorted one. In addition, current NMR spectrometers use amplifiers to convert the radiofrequency (RF) to audio-frequencies (AF), by carrying out the subtraction of the reference frequency from the detected signal (Sect. 1.4), which are easier to handle (Fig. 1.7). An ideal amplifier would be a wire with a high-gain, located adjacent to the probe to reduce signal losses due to the connecting cables. The signal is converted from analogue to digital by an analogue-to-digital converter (ADC or digitizer). It is necessary to use an ADC because the signal from the probe is recorded in analogue form (since the FID is the superposition of various frequencies), but it must be in a digital form to apply the FT in the computer. The output signal of the amplifier must match the dynamic range of the ADC. To faithfully reproduce the analogue signal, the sampling rate of the ADC must be at least twice the frequency of the signal (*Nyquist condition*). Even if the Nyquist condition is fulfilled, the ADC makes an approximation to the true signal, since it represents the superposition of the frequencies by 16 to 18 bits (that is, the most intense signal can be digitized as 2^{18}). This can have serious consequences for the dynamic range (the limits on weak and strong signals we can observe): for instance, signals with less than $1/2^{18}$ of the signal amplitude cannot be distinguished. Errors in the digitization of strong signals can originate "noise" on the baseline, which may "swamp" weak signals. Therefore, the receiver gain (the amplification of the signal) should be adjusted in such a way that the most intense signal completely fills the range of the amplifier.

1.5.3 The Lock-System

The frequencies generated in modern NMR spectrometers are very stable, but the Larmor frequencies ($\omega_0 = \gamma B_0$) of the nuclei are not. Since there is no inherent instability in the constant gyromagnetic ratio (γ) of the nuclei, the variability must come from the static magnetic field B_0. The field created by the superconducting magnet decays slowly with time and it is not stable enough for high-resolution NMR. To improve the stability of the field, an electronic device has been incorporated into modern NMR spectrometers tasked with the acquisition of a second NMR experiment in a continuous and automatic manner. This experiment uses deuterium (^2H) as the monitored nucleus, which is why the inner coils of the probes are always tuned to this nucleus. To ensure the presence of the ^2H isotope in reasonable amounts in the sample, 5–10 % of ^2H_2O is added to aqueous samples or fully deuterated organic solvents, like $CDCl_3$ or DMSO-d_6, are used.

This second, stabilising experiment consists of a continuous train of pulses applied to the sample at the Larmor frequency of deuterium, which will vary depending on the solvent used. The Larmor frequency of the deuterons is established by bringing the signal onto resonance through the manual adjustment of the field. The electronic circuit of the device maintains the on-resonance condition, locking the field to the exciting frequency of deuterium; however, since all frequencies in modern spectrometers are generated from the same source, all the excitation frequencies for any nuclei explored in the experiment are locked. The system works by using a feedback loop, which generates corrections to the B_0, such that the on-resonance frequency of the deuterium in the solvent remains constant.

1.5.4 The Transmitter/Receiver System: Quadrature Detection

We know from trigonometry that the values of $\cos(\alpha)$ and $\cos(-\alpha)$ are equal. Therefore, since we are registering the decay of the transverse magnetization after a radiofrequency pulse with a single receiver coil located along one of the axis in the xy-plane (Fig. 1.9a), we cannot distinguish between frequencies rotating slower or faster than the frequency of reference (*carrier frequency*), that is, negative or positive frequencies, respectively.

One possible solution would consist in setting up the carrier frequency on one side of the spectrum instead of the centre, but this would increase the noise in the spectrum, which would be only averaged on "one" side of the reference frequency. Another solution to distinguish the positive from the negative frequencies would be to arrange a second detector along the other axis (the x-axis in our discussion, which is shifted 90° with respect to the y-axis); then, by collecting the sine projection of the transverse magnetization it will be possible to differentiate

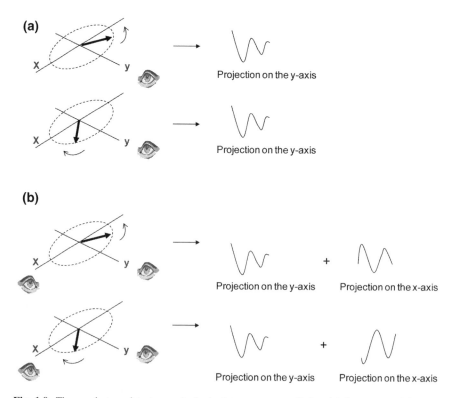

Fig. 1.9 The quadrature detector. **a** A single detector cannot distinguish between positive and negative frequencies. **b** Two detectors located in quadrature (that is, they are shifted by 90°) can discriminate between the sense of precession of two signals

the sign of the frequencies (Fig. 1.9b). This approach is known as *quadrature detection* since both detectors are shifted by a quadrant (90°) of a circle. However, in a cramped system like an NMR probe, the introduction of a second coil is difficult. Therefore, the sine- and cosine-modulated components of the signal come from a different trick: the signal coming from the receiver coil is divided into two signals by a splitter. Both parts are mixed with the transmitter frequency (the carrier frequency). The phase of the transmitter frequency that is added to one of the split signals differs by 90° from that of the other half, and therefore, we shall have the equivalent of having two detectors. Experimentally, sine and cosine-modulated signals are sometimes amplified slightly different, and artefacts called *quadrature images* can appear. These artefacts can be recognized because they are "mirrored" about the zero frequency (the middle of spectrum, where the carrier frequency is located). To avoid these undesired effects, modern spectrometers utilize oversampling, where a larger spectral width is used such that the quadrature images do not fall into the observed spectrum.

1.5.5 The Shim System

Present generation magnets have field inhomogeneities of 1 part in 10^9, thanks mainly to the careful design of the internal solenoid, although it can still be considered an insufficient value for the acquisition of high resolution NMR experiments. The shim system consists in a hardware device that corrects for slight differences in the local magnetic fields. It is constituted of two parts: the cryo-shim system and the room-temperature shims. The principle behind both is alike: small coils are supplied with highly regulated electrical currents. These currents produce small magnetic fields that correct the field inhomogeneities created by the main solenoid. There are several of those coils positioned at diverse geometries to produce field corrections with different orientations. The cryo-shim system is located inside the He bath and after the initial adjustment by the engineers during the start-up of the instrument, no further manipulation is required. The room-temperature shim system is set coaxially in the bore of the magnet, it is user-controlled and needs adjusting every time a new sample is placed in the instrument. The room-temperature systems (of which there are typically 20 to 30 on a modern instrument) are usually grouped into two classes: the on-axis spinning systems (z, z^2, z^3, z^4) and the off-axis non-spinning shims (x, y, xy, and others). The *on-axis* shims only correct for field inhomogeneities along the z-axis, being highly dependent on the solvent and the height to which the NMR tube is filled. The *off-axis* shims normally do not have to be extensively adjusted, and only x, y, xz and yz need some tinkering with for every new sample. In general, shimming is only practiced on a small fraction of the total available number of shims. Shimming improvement is usually performed by either observing the intensity of the lock signal (which one aims to maximize in height) or by monitoring the shape of the FID. However, the recent introduction of the automation of the process of shimming has eased this pre-acquisition step considerably.

1.5.6 Pulse Field Gradients

One of the most important technical advances in high-resolution NMR during the last decade of the twentieth century was the introduction of *pulse field gradients* (PFG), or *gradients* for short. In general, gradients are used to diffuse unwanted signals and thereby sharpen the desired ones (see below). Gradients had been applied in magnetic resonance imaging for a long time, but their application to high-resolution NMR was limited by technical problems (Hurd 1990). Among these difficulties was the production of *eddy-currents* generated during the application of the gradient pulse. These are currents generated in the components that surround the sample, like the circuitry and the shims themselves, which in turn generate magnetic fields within the sample. Since these currents can last for hundreds of milliseconds, acquisition of high-resolution spectra was hampered under their influence. The most effective way to suppress the generation of *eddy-currents* (and therefore, the solution to the technical problem) is the use of actively-shielded gradient coils, currently used in all commercially available gradient probes. This extra gradient coil generates a second field outside the active sample region, opposing and cancelling out the field produced by the inner coil. In the next paragraph, we shall describe the physical basis of gradients and how they are produced. Any reader interested in the more theoretical aspects and applications of pulse field gradients can find a more detailed description in recent literature (Keeler et al. 1994).

1.5.6.1 Gradients

If we imagine the NMR sample as divided into small disks, after a radiofrequency pulse each disk will have a magnetization vector in the transversal xy-plane, but each of them will experience a slightly different local magnetic field according to its position in the sample (Fig. 1.10, left). If a linear variation of a magnetic field is imposed for a short time (a gradient pulse) along the z-axis (grey short arrows), chemically equivalent spins will experience different fields according to their position along the z-axis, and therefore they will precess with diverse frequencies and rotate at dissimilar angles. If the gradient is long and powerful enough, the sum of all transverse magnetizations will be null and there will be no detectable NMR signal (since the signal is highly inhomogeneous). The gradient will have completely defocused the magnetization. However, although the magnetization is globally null it is still present in the sample, and the application of a second gradient (with the appropriate power, length and phase) will undo the dephasing achieved with the first gradient yielding a refocusing of all phases in the imaginary

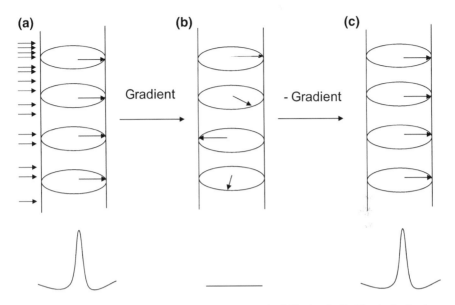

Fig. 1.10 How PFGs work. **a** In a homogenous magnetic field, chemically identical spins have the same phase and Larmor frequency after the application of a (90°)-pulse, yielding a single sharp resonance (*bottom*). The *grey arrows* on the *left side* indicate the field variation, that is, the gradient (with smaller intensity at the *bottom* of the figure and higher at the top). **b** After the application of a linear gradient, the spins have adopted different phases and no signal is detected. **c** Application of a second identical gradient with the opposite direction refocuses the individual spins resulting in a re-phased peak

sample disks, resulting in observable transversal magnetization (Fig. 1.10, right). The concept of defocusing the signals and refocusing only the desired ones (by modulating the power, the length and the phase of the refocusing gradient) has made PFGs extremely useful in NMR. In fact, conversely to what happens with radiofrequency pulses, accurate strength and pulse calibration of PFGs are not required for experimental success, provided that the PFG is strong enough to dephase unwanted magnetization. However, it is important to ensure that the PFGs are reproducible and that the gradient ratios of the pulse gradients used to defocus and refocus are kept constant.

Gradients are used in high-resolution NMR in three general applications. First, the removing of unwanted magnetization such as the solvent signal, which is of widespread use in aqueous biomolecular samples. PFGs also eliminate responses resulting from imperfections of pulse sequences, an important application that replaces the use of time consuming phase cycles (Sect. 1.4.2). Second, PFGs are used to select magnetization routes crucial to the NMR experiment while discarding non-relevant ones. We shall not describe this application, which is beyond the scope of this book [the interested reader can find more information in (Claridge 1999; Cavanagh et al. 1996; Keeler et al. 1994)]. And finally, gradients are used in shimming (Vanzijl et al. 1994; Barjat et al. 1997), where inhomogeneities of the magnet due to the sample are encoded in the applied gradients (as phase differences, Fig. 1.10) and used to correct them.

Regarding gradient generation, they can be produced by offsetting one of the shim currents from its optimum value (for instance, driving the z-shim at its maximum current), although such set-up only creates weak fields and offers no control over the gradient amplitude. In current probes, there are dedicated-gradient coils surrounding the usual radiofrequency solenoids, together with an appropriate gradient amplifier and a second shield-gradient coil (Sect. 1.5.2). The field gradient is generated by applying a current to the gradient coils (the typical highest value is 0.5 Tm^{-1} or $50\,Gcm^{-1}$) or by reversing it, if an inversion of the gradient phase is sought.

Finally, the use of gradients might appear incompatible with locking the field, since gradients introduce inhomogeneities which would preclude the electronics from keeping the ^2H lock signal fixed. However, the corrections introduced by the lock are the result of long-time monitoring periods of the static field, while the perturbations caused by gradients only last a few milliseconds at most. In experiments with gradient application, the lock should be highly "damped" (in a sense, disconnected) to avoid the computer trying to correct the field each time is recovering from the effect of an applied gradient.

1.5.7 Sample Preparation

Most NMR experiments in the biology/biochemistry field are carried out in H_2O, with the addition of 5–10 % of ^2H_2O for the lock signal. In some cases, especially when working with peptides or protein fragments, organic solvents such as trifluoroethanol or dimethyl sulfoxide are used. For experiments with proteins associated with lipid membranes, the use of detergents such as sodium dodecyl sulphate (SDS) is required, and in some cases, conventional high-resolution NMR methods cannot be used; an alternative to gain structural insight into those systems is offered by special techniques like solid-state NMR. Organic solvents are usually employed in chemistry, most of them commercially available in deuterated forms to allow acquisition of ^1H-NMR spectra without interference from the solvent signal.

As NMR is an intrinsically insensitive spectroscopic technique, experiments aimed at structural molecular elucidation tend to be performed at high sample concentrations (in the mM–M range). Therefore, the molecule must be soluble at those concentrations (i.e. monodisperse), and be stable enough to last during the usually long acquisition times. In the case of stable biomolecules, NMR experiments are normally acquired at pH ≤ 7.0, since under these conditions the hydrogen-exchange reaction of amide protons with water is minimised. Buffer components presenting active spins that could obscure any of the resonances of the sample are to be avoided, although deuterated chemicals can be employed in homonuclear ^1H-detected experiments. There is no need for the use of deuterated solvents for the acquisition of NMR experiments observing other nuclei rather than ^1H (e.g. ^{13}C, ^{15}N). High ionic-strength samples decrease the signal-to-noise ratio

(S/N) as the presence of local magnetic fields caused by the ions produce an increase in pulse lengths. Finally, as the resonance line width decreases (and therefore, resolution increases) with temperature, NMR experiments are normally acquired at room temperature or higher, provided that the molecule does not suffer any conformational changes or degradation. If a system under study requires low temperatures, NMR experiments can be carried out even below 173 K by keeping the sample cold with a continuous flow of liquid N_2 and using organic solvents with low freezing points (CH_2Cl_2, tetrahydrofuran (THF) or others).

References

Barjat H, Chilvers PB, Fetler BK, Horne TJ, Morris GA (1997) A practical method for automated shimming with normal spectrometer hardware. J Magn Reson 125:197–201

Berger S, Braun S (2003) 200 and more NMR experiments. Wiley-VCH, New York

Bracewell RM (1978) The Fourier transform and its applications, 2nd edn. McGraw Hill, New York

Cavanagh J, Fairbrother WJ, Palmer AG 3rd, Skelton NJ (1996) Protein NMR spectroscopy. Principles and practice, 1st edn. Academic Press, New York

Claridge TDW (1999) High-resolution NMR techniques in organic chemistry. Pergamon Press, Oxford

Derome AE (1987) Modern NMR techniques for chemistry research. Pergamon Press, Oxford

Ernst RR, Bodenhausen G, Wokau A (1987) Principles of NMR in one or two dimensions. Clarendon Press, Oxford

Evans JNS (1996) Biomolecular NMR spectroscopy. Oxford University Press, Oxford

Farrar TC, Becker ED (1971) Pulse and Fourier transform NMR. Academic Press, New York

Günther H (1995) NMR spectroscopy: basic principles, concepts and applications in chemistry, 2nd edn. Wiley, Chichester

Harris RK, Mann BE (1978) NMR and the periodic table. Academic Press, London

Hurd RE (1990) Gradient enhanced spectroscopy. J Magn Reson 87:422–428

Keeler J (2006) Understanding NMR spectroscopy, 2nd edn. Wiley, Chichester

Keeler J, Clowes RT, Davis AL, Laue ED (1994) Pulse-field gradients: theory and practice. Methods Enzymol 239:145–207

Mason J (1987) Multinuclear NMR. Plenum, New York

Sanders JKM, Hunter BK (1992) Modern NMR spectroscopy, 2nd edn. Oxford University Press, Oxford

Vanzijl PCM, Sukumar S, Johnson MO, Webb P, Hurd RE (1994) Optimized shimming for high resolution NMR using three-dimensional image-based field mapping. J Magn Reson 111:203–207

Wüthrich K (1986) NMR of proteins and nucleic acids. Wiley, New York

Chapter 2
Spectroscopic Parameters in Nuclear Magnetic Resonance

Abstract In common with other spectroscopic techniques, NMR measures the intensities of the nuclear spin transitions versus the frequencies at which these transitions happen. In this chapter, the spectroscopic parameters of NMR will be described, some of which are shared by other spectroscopies while others are unique to the technique, arising from the properties described in the previous chapter. The origin and physical characteristics of the chemical shift, the spin–spin scalar coupling and the nuclear Overhauser effect (NOE) will be introduced here.

Keywords Chemical shift · Coupling constant · Multiplets · Nuclear Overhauser Effect (NOE) · Relaxation

2.1 The Chemical Shift and the Spectral Intensity

The NMR frequency of any particular nucleus is *spectrometer dependent*, that is, carries with it the effect of the static field applied (B_0), which is the origin of the split levels that generate the NMR effect (Chap. 1). In order to be able to compare spectra recorded at different field strengths we need to define other parameters. The chemical shift parameter (δ) is central to the interpretation of NMR spectra, and it is equivalent in importance to the frequency, wavenumber or wavelength used in other spectroscopies such as IR, CD, UV absorbance or fluorescence.

2.1.1 The Shielding Screening Constant

The basic equation in NMR described in Chap. 1, $\omega = \gamma B_0$, tells us that when any type of nucleus in the sample is subjected to the same external magnetic field, B_0, it will experience the corresponding Larmor frequency (according to its γ). Therefore, the NMR spectrum will be composed of identical resonance lines of the

R. J. Carbajo and J. L. Neira, *NMR for Chemists and Biologists*,
SpringerBriefs in Biochemistry and Molecular Biology,
DOI: 10.1007/978-94-007-6976-2_2, © The Author(s) 2013

same frequency for a particular type of nucleus. Fortunately for our purposes, this
is not the case experimentally and each nucleus resonates with a particular frequency. To show this point, we consider as an example a simple molecule such as
ethanol (CH$_3$CH$_2$OH) that would present an "ideal" proton NMR spectrum as that
shown in Fig. 2.1 (main panel). Three groups of signals can be clearly differen-
tiated (their splitting is discussed later in this chapter), corresponding to the
chemical differences of the hydrogen atoms in the molecule: one group of signals
corresponds to the methyl group (-CH$_3$), a second one to the methylene protons
(-CH$_2$-) and a third for the hydroxyl group (-OH). Chemically speaking, it is
obvious that each group of hydrogens in the ethanol molecule has different
molecular environments, and therefore will experience slightly different local
fields. If we compare the frequency for each ^{1}H nucleus in ethanol with that
calculated by γB_0, we find that each resonance frequency measured experimentally
is different from the theoretically predicted one. That is, any of the resonance
frequencies observed can be described as:

$$\omega = \gamma B_0 - \sigma\gamma B_0 = \gamma B_0(1 - \sigma)$$

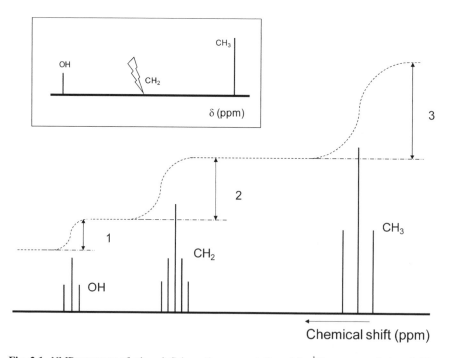

Fig. 2.1 NMR spectrum of ethanol. Schematic representation of the ^{1}H spectrum of ethanol. The
dotted line indicates the integral of the signals which correlates with the number of equivalent
protons. Inset: spectrum of ethanol when the CH$_2$ resonance is irradiated (the bolt) causing the
decoupling of the rest of signals (Sect. 2.2.4)

where the factor σ is called the *shielding* (or *screening*) *constant*. σ can be positive or negative depending on whether the local field adds or subtracts from the applied field. The σ constant varies with B_0 along the axes (x, y and z) within the molecule, and can be described mathematically as a tensor. Since σ varies with the chemical environment, identical atoms within a molecule (i.e. the different protons in ethanol) will resonate at different frequencies and therefore render unique NMR signals, giving rise to the richness and power of NMR spectroscopy.

It is not possible to calculate the precise value of σ for any given nucleus in any molecule due to the different mechanisms at work. However, we can try to understand the variation within molecules by considering the different types of effects that contribute to it:

$$\sigma = \sigma_d + \sigma_p + \sigma_m + \sigma_{rc} + \sigma_{ef} + \sigma_{sol} + \sigma_{ps}$$

The contributions are from local diamagnetism (σ_d), local paramagnetism (σ_p), neighbouring anisotropy (σ_m), ring-currents (σ_{rc}), electric field (σ_{ef}), solvent contribution (σ_{sol}) and the effect due to the presence of paramagnetic species (σ_{ps}). The origin and effect of each one of these screening constant contributions is described in the following.

2.1.1.1   The Diamagnetic Contribution

Every atomic nucleus is surrounded by an electron cloud (except H^+). The presence of an external B_0 causes the electrons to precess about the axis of the field, creating a current that gives rise to its own small magnetic field (Fig. 2.2a). The σ_d contribution can be calculated from the value of the scalar product of the electron-nucleus distance and the electron orbital density. The diamagnetic effect describes the behaviour of spherically distributed electronic clouds, such as the s-orbital in protons. Therefore, its effect is particularly important in hydrogen atoms (so called *protons* or ^1H in NMR jargon), but not that relevant in heavier nuclei. It is also affected by the presence of other nuclei withdrawing or donating electronic charge, that is, by alterations in the electron clouds. Thus, highly electronegative nuclei, like oxygen or nitrogen, will withdraw electronic charge, with the effect of deshielding nearby hydrogen atoms. A case in point is the downfield (deshielded) resonance of the ^1H at the C1 position in glucose due to the high electronegativity of the nearby ring oxygen.

Even in hydrogen, the distribution of electrons is usually non-spherical, and the shielding depends on the orientation with respect to B_0. In sp^3 carbons, protons are tetrahedrally bonded and the electron distribution is invariable upon rotation. On the other hand, for sp^2 carbons (e.g. C=C, C=O groups) the shielding depends highly on the orientation of the bond relative to B_0. Due to this anisotropy, the effective field that the nuclei actually experience can deviate with the direction of B_0, and different directions of the field will result in different resonance positions for the same chemical species. This effect is known as *chemical shift anisotropy*

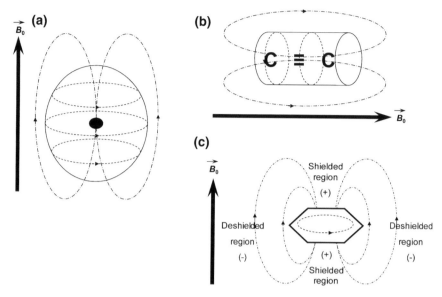

Fig. 2.2 Contributions and effects of the shielding parameter. **a** Diamagnetic effect: the *dashed lines* indicate the movement of the electron within its orbital (large continuous circle); the *dot-and-dashed lines* indicate the magnetic field created by the electronic current; the *filled circle* represents the nucleus. The direction of the external magnetic field is shown by an arrow on the left-side of the figure. **b** Magnetic anisotropy of a triple bond: only the carbons of the triple bond are shown. The *dashed lines* indicate the movement of the electrons within the triple bond; the *dot-and-dashed lines* indicate the field created by the electronic current of the triple bond. The direction of the external magnetic field is indicated by an *arrow* at the bottom of the figure. **c** Ring-current effect: the *dot-and-dashed lines* indicate the direction of the magnetic field created by the ring-current. The dashed line on the ring shows the movement of the electrons in the aromatic plane. Shielded and deshielded regions are indicated. The external magnetic field is indicated by an *arrow* on the *left* side of the figure (this panel has been adapted from Structural Virology book by Springer, with permission)

(CSA). In solids, where molecular rotation is hindered, the line width of each resonance is significantly broadened due to the CSA. In liquids, where rotation is not hampered, CSA contributes to the relaxation of spins (Chap. 1) and is, in fact, the major relaxation mechanism for non-protonated carbons.

2.1.1.2 The Paramagnetic Contribution

The circulation of electrons around p-orbitals induced by external magnetic fields causes deshielding (downfield shifts) of the nearby nuclei. The paramagnetic term can be calculated if the way in which the electron wave-functions are altered by the presence of the other nuclei is known. The shifts caused by the paramagnetic contribution are much larger than for the diamagnetic effect, and this contribution is the main source for chemical shift variation in all nuclei heavier than ^{1}H.

2.1.1.3 Neighbour Anisotropy and Ring-Current Effect Contributions

Some types of chemical bonds create an additional magnetic field which is anisotropic in space. A typical example is the carbon-carbon triple bond $C \equiv C$ (Fig. 2.2b), where the π electrons form an electron cloud extending around the bond axis in the shape of a cylinder. The B_0 forces the π electrons to rotate around the bond axis creating a magnetic field, which counteracts the static magnetic field.

The same effect is observed in the π-cloud of aromatic systems. In the presence of B_0, a planar aromatic ring with its delocalized π electrons "creates" its own magnetic field (Fig. 2.2c). A nucleus below or above the ring plane will be shielded from the external magnetic field, but it will be deshielded if it is located at other positions relative to the ring. Therefore, the magnitude of the total field experienced by a nucleus informs us on its orientation with respect to the ring. Such anisotropies can dramatically alter the shape of spectra, increasing the spread of signals across the chemical shift range. For instance in the case of proteins, the effects of the aromatic rings from amino acids like histidine, phenylalanine, tyrosine and tryptophan, can be used as indicators for the presence of protein structure: well-spread resonances are a sign of the close proximity of those aromatic rings to other residues, and consequently of a folded protein.

2.1.1.4 The Electric Field Contribution

Strongly polar groups create intramolecular electric fields, which distort the electron density in the rest of the molecule, influencing σ. The electric field effects account for the electron movement in the polar bond against the inherent electric field, and the asymmetry induced by this electronic shift. Electric fields are important, for example, in ^{19}F resonances as well as in hydrogen-bonded protons. Hydrogen-bonds decrease the electron density at nearby chemical bonds, and lead to a high frequency resonance. In fact, hydrogen-bonded protons can be easily recognized by their shifted frequency, although this is highly dependent on experimental sample conditions like temperature, concentration and solvent.

2.1.1.5 The Solvent Contribution

The effects of solvents on the spectra of solutes can be manifested through several mechanisms. In the first place, aromatic solvents such as benzene or toluene will possess ring-current effects like those described before. Second, an orientation effect can be found when a polar molecule is dissolved in a polar solvent, producing an ordering of solvent molecules around the solute that alters the magnetic field experienced by the latter. A final effect is encountered when hydrogen-bonds are formed between the solute and the solvent.

2.1.1.6 The Contribution of Paramagnetic Compounds

Paramagnetic compounds are those that present unpaired electrons. Paramagnetic entities in an NMR sample give rise to signal broadening due to relaxation effects. The resonance frequency can also be influenced through Fermi-contact interactions (Sect. 2.2.1) between a nuclear spin and an unpaired electron from the paramagnetic species. There are several examples in the literature describing the use of paramagnetic compounds in NMR. For instance, the so-called *shift-reagents* are paramagnetic chemicals used with the purpose of dispersing previously overlapped resonances in order to clarify proton spectra (e.g. Cr(III) in Cr(acac)$_3$). The paramagnetic contribution is also observed in the NMR spectra of some metal nuclei. In this type of atoms, the dipolar coupling between the nucleus and the electron can cause signal broadening, but if the coupling is much smaller that the relaxation rate of the electron, sharp lines with large resonance frequencies are observed in the NMR spectrum of the metal (*Knight shift*).

2.1.2 The Chemical Shift

To quantify the resonance frequencies, we need signal reference standards. The common standard for protons is the ^{1}H resonance of Si(CH$_3$)$_4$ (tetramethylsilane, abbreviated as TMS), a chemically stable molecule that presents a unique ^{1}H resonance, which is very intense as it is contributed to by 12 equivalent proton atoms, to which the arbitrary value of 0 Hz frequency is assigned. Conveniently, the zero reference frequency for the ^{13}C nucleus is also that of the methyl carbon of TMS. Alternatively, the ^{1}H signal of the residual non-deuterated solvent in the sample (which has been previously referenced to TMS) can also be used as reference, like for instance CHCl$_3$ or DMSO ((CH$_3$)$_2$SO). For the H$_2$O/D$_2$O samples typically used in biomolecular NMR, ^{1}H resonances are normally referenced to (4,4-dimethyl-4-silapentane-1-sulfonic acid) DSS or (trimethylsilyl propionate) TSP, both resonating at 0 Hz. Each nucleus in the periodic table has its own chemical as NMR standard, although in biomolecules the common procedure to calibrate the ^{13}C and ^{15}N nuclei is to take into account the reference of ^{1}H (Wishart et al. 1995).

Even though a reference is used to calibrate the signals in the spectra, the separation in hertz of a particular resonance from the standard signal depends on the B_0 at which they are recorded, as the higher the field the greater the separation between the energy levels and therefore the frequency of absorption (Chap. 1). To be consistent at any B_0, we define *chemical shifts*. The chemical shifts are dimensionless units reported on the δ scale (in ppm), which is defined as $\delta = ((\upsilon - \upsilon_0)/\upsilon_0) \times 10^6$, where υ_0 is the resonance frequency of the standard, υ is the frequency of the particular nucleus, and the 10^6 factor gives the scale of the magnitude in parts per million. NMR spectra are represented with the zero δ in the right-hand side of the frequency range, with increasing values towards the left side, an old remnant of early NMR experiments when spectra were acquired by continuously varying B_0. If $\delta < 0$ the nucleus is considered shielded because the magnetic field it experiences is weaker than the field experienced by the nuclei in the standard compound. However, if $\delta > 0$, we say that the nucleus is deshielded because the local magnetic field is stronger than that experienced by the nuclei in the standard molecule under the same conditions. In general, the closer the nucleus is to an electronegative element, the more deshielded it will be (i.e. the larger its δ value). The values of δ can be used to make qualitative assessments about the presence of functional groups, in a similar fashion as functional group frequencies are identified in IR spectroscopy.

Based on the above arguments, several general statements can be made for interpreting proton spectra. Thus, in the aliphatic C–H bonds the shielding decreases in the series $CH_3 > CH_2 > CH$, with the methyl groups resonating around 0.9 ppm, while those corresponding to CH protons lay between 3.5 and 10.0 ppm (Fig. 2.3). There are some exceptions to this general rule, which occur when the observed proton is attached to an atom or group of atoms presenting high electronegativity. Aromatic protons appear between 6.0 and 8.0 ppm because, in addition to their sp^2 hybridization, an extra deshielding contribution arises due to the ring-effects (Sect. 2.1.1.3). The resonance signals of protons belonging to aldehydes and carboxylic acids appear at very large δ values (>8.5 ppm). Carbon atoms bound to nitrogen, oxygen or halogens (very electronegative elements)

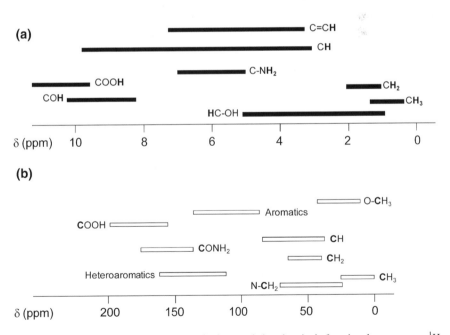

Fig. 2.3 Resonance frequency ranges of characteristic chemical functional groups. **a** ^1H chemical shifts. **b** ^{13}C chemical shifts

induce large displacements in the resonance of neighbouring protons leading to deshielded chemical shifts (therefore, larger values). The NMR signals of labile protons like OH, NH, NH_2 and CO_2H are concentration-, temperature- and solvent-dependent due to possible exchange reactions with the solvent (Fig. 2.3a) (Günther 1995).

These chemical shift rules can also be applied to ^{13}C. For instance, the chemical shifts of ^{13}C belonging to aliphatic chains are shielded (low δ), and those involved in aldehydes or acids are deshielded. However, since the γ of ^{13}C is different to that of ^1H (Table 1.2), the range of frequencies/chemical shifts is very different, with ^{13}C δs extending to above 200 ppm (Fig. 2.3b) (Günther 1995).

2.1.3 Signal Intensity

Each NMR signal encloses an area under the peak that can be quantified. As in any other spectroscopy, this area under the spectral curve correlates with the number of molecules or nuclei originating that transition. Together with the chemical shift information, we can use the relative intensities under each peak to determine which nucleus, or group of nuclei, give rise to a particular resonance; that is, we obtain structural and chemical characterization.

As an example, we shall use again the ethanol molecule. If we integrate with the adequate software the area under each peak, we should obtain relative intensities in the ratio 3:2:1 (Fig. 2.1, dashed line). The signal appearing more shielded (lowest δ) with a relative intensity of three correlates with the methyl (CH_3) group. That with the relative intensity of two correlates with the CH_2 group and is deshielded compared to the methyl one due to the electronegative effect of the bound oxygen atom. The remaining signal integrating for one ^1H shows the largest δ and corresponds to the O–H group.

2.2 The Scalar Coupling Constant

Any interaction that divides magnetically equivalent nuclei into two or more populations will cause its peaks in the NMR spectra to split. *Magnetic equivalent* nuclei are those presenting the same frequency (isochronous) and also identically coupled to any third nucleus. For instance, the protons of a methyl group are chemically and magnetically equivalent as a consequence of the free rotation about the C–C bond. All three protons have the same time-averaged chemical environment and hence the same frequencies. In contrast, *chemical equivalent* nuclei behave the same as one another chemically, they are also isochronous, but they can present different magnetic interactions to a third nucleus. All magnetic equivalent nuclei are also chemically equivalent, but the opposite is not necessarily true.

Two types of interaction between spins are known: *dipolar* and *scalar* coupling. The dipolar coupling depends on the orientation of the connecting vector between the two nuclei with respect to the static field (Chap. 1, Fig. 1.3a); this orientation changes rapidly in solution due to molecular tumbling, averaging the dipolar coupling to zero in isotropic (liquid) phase (although it can be observed in solid state and in liquid crystals). Therefore, it does not cause any possible splitting on the resonances in the solution state. On the other hand, the scalar coupling does not depend on molecular orientation but is transmitted through chemical bonds, inducing the splitting of resonances. This coupling will be described in detail in the following sections.

2.2.1 Spin–Spin Coupling

If we go back to the ethanol molecule (Fig. 2.1, main panel), its ^{1}H-NMR spectrum shows splitting in its three resonances: CH_3, CH_2 and OH. Instead of a single sharp line for each of the three signals, the CH_3 resonance appears as a triplet, as does the OH and the CH_2 as a quintuplet. The fine structure of the NMR spectrum arises from the fact that each magnetic nucleus contributes to the local field experienced by its neighbours, slightly modifying their resonance frequencies. The strength of this interaction is proportional to the scalar product of the magnetic moments of both interacting nuclei, with a proportionality constant called the *coupling constant*, usually abbreviated to ^nJ (n indicating the number of intervening covalent bonds). The J coupling is measured in hertz, is governed only by the magnetic moments of the nuclei involved, and is independent of the applied B_0 (i.e. J will be identical for two coupled nuclei regardless of the NMR spectrometer used). The coupling is transmitted via the chemical bonds and, in general, its magnitude decreases with the distance (number of bonds) between the coupled nuclei. The strength of J also depends on the amount of overlap of the involved electron orbitals (Sect. 2.2.2).

To understand the effect of scalar coupling on NMR signals, consider a molecule that contains two spin-½ nuclei, A and X (Fig. 2.4). In the absence of scalar coupling, we have four levels depending on the orientations relative to the external magnetic field of the two spins: both oriented parallel or antiparallel, or one parallel and the other antiparallel, and the other way round. The transitions expected for the A nucleus (going from the α state to the β state, with the X spin in either α or β states) are drawn in Fig. 2.4a. The A nucleus will resonate at a certain frequency as a result of the external field and its inherent σ, giving rise to a single line for those two transitions. If there is a spin–spin coupling between the nuclei, the two states with both A and X spins presenting the same orientation (that is, either both nuclei in the α or β state) will be destabilized (they will have more energy in the presence of the spin–spin coupling interaction), and the two states with the two A and X spins in alternate orientations will be stabilized. The final result will be that the two transitions for the A spin will have different energies and

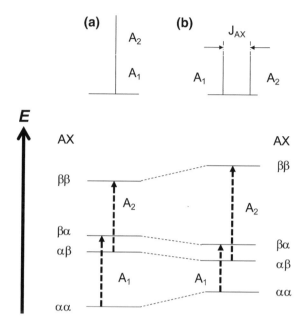

Fig. 2.4 Nuclear energy levels for a two-spin system with spin ½. Energy diagram without spin–spin coupling **a** and with coupling **b**. The parallel orientation of both spins was considered as the low-energy state. The levels $\alpha\beta$ and $\beta\alpha$ have the same energy (indistinguishable from a quantum perspective), but they are shown here with slightly different energies for the sake of distinguishing the A_1 and A_2 transitions. The A_1 transition changes the state of the A nucleus from α to β when the X state is α; the A_2 transition changes the state of the A nucleus from the α to β when X state is β. Therefore, in the spectrum for the A resonance, in the absence of coupling, a single peak appears corresponding to the sum of A_1 and A_2 (above on the *left* side). Note that the lengths of the *arrows* indicating the transitions A_1 and A_2 have equal length, because $\alpha\beta$ and $\beta\alpha$ are degenerate. In the presence of coupling, two peaks (A_1 and A_2), separated by J, correspond with the A nucleus (above on the *right* side). The same splitting would be observed for the X nucleus (not shown)

two lines will appear instead of the previous single resonance (Fig. 2.4b). The separation (in Hz) between the two lines forming a doublet is defined as the magnitude of the coupling J_{AX}. The same effect will be observed for the coupled spin X: two lines will appear instead of one, equally separated by J_{AX} Hz.

If a second A nucleus which is magnetically and chemically equivalent to the first (with the same chemical shift) is present in the molecule (Fig. 2.5a, (1)), the X resonance signal will be split into a doublet by the first A nucleus and each of the doublet signals split again by the second A nucleus (Fig. 2.5b, (1) and (2)). Because the A nuclei are magnetically equivalent, the size of the J_{AX} coupling will be identical, and the doublets will add up to yield a triplet for the X resonance with an intensity ratio 1:2:1 (Fig. 2.5b, (1)). In the case of three magnetically equivalent X nuclei coupled to a single A (Fig. 2.5a, (1)), following the same reasoning above the A resonance will be split into four lines of intensity 1:3:3:1 (Fig. 2.5b, (1) and

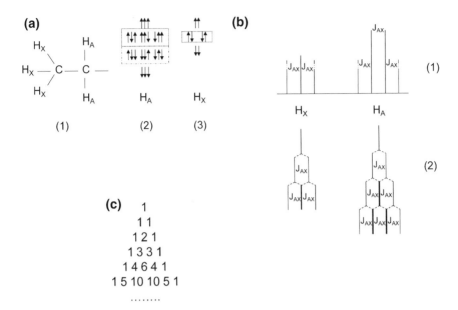

Fig. 2.5 Splitting in a system with spin-½ nuclei. **a** *Left* (1): scheme of the molecule with three equivalent H_X nuclei and two equivalent H_A nuclei. *Middle* (2): the different possible orientations for the H_A due to the presence of H_X ($2^3 = 8$). The orientations marked with a dotted square on one side, and those with a dash-and-dot square on the other are equivalent, because the spin orientations within each set of spins within a square are indistinguishable. *Right* (3): possible orientations for the H_X caused by the presence of H_A ($2^2 = 4$). The orientations marked with a *dotted* square are equivalent, because the spins are indistinguishable. **b** *Top* (1): the resulting spectra for each of the H_A and H_X nuclei are shown. *Bottom* (2): diagram showing how the lines of the multiplets above are generated. Lines with different width indicate the overlapping of two or more lines of the multiplet. The separation between the lines corresponds to the coupling J_{AX}. **c** The Pascal triangle provides the intensity of each of the lines in the multiplet. Each number in a line of the triangle lines is the result of the addition of the numbers immediately above it

(2)), with the quantum states with several nuclei in alternate orientations having identical occupations (Fig. 2.5a, (2)). The general rule is that N magnetically equivalent spin-½ nuclei will split the resonance of a nearby spin (or group of spins) into $N+1$ lines with a distribution intensity given by the Pascal triangle (Fig. 2.5c). If the spins involved do not have $I = $ ½, then the number of split lines is $2IN + 1$. It is important to note that there is no coupling among the three H_X since they are magnetically equivalent (and the same is true for the two H_A).

The above rules for signal multiplicities only apply for so-called *weak coupling*. This occurs when the chemical shift difference between the coupled nuclei is very large compared to J (Fig. 2.6, top spectrum). In the case of weak coupling, the flipping of one spin does not result in a tendency of the other spin to flip, and discrete resonance lines can be assigned to one or the other. The interested reader can look into this topic in more advanced NMR books or reviews (Günther 1995; Sternhell 1969; Barfield et al. 1990; Homans 1989). When the chemical shift

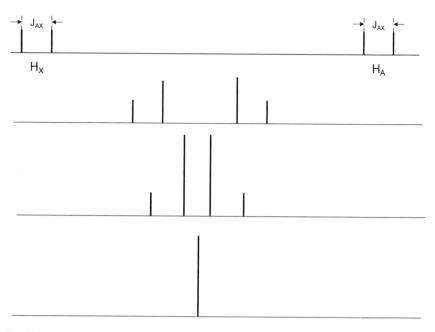

Fig. 2.6 From weak to strong coupling. In the *top* spectrum the chemical shift separation between the A and X nuclei is much larger than the J_{AX}. In the second spectrum from the *top*, the J_{AX} is half the size of the chemical shift separation between the two nuclei. In the third spectrum, the J_{AX} and the chemical shifts are very similar. For magnetically equivalent nuclei (*bottom* spectrum) only one line with double intensity is observed. As the difference in chemical shift decreases, the inner lines increase in intensity while the J_{AX} is kept constant (the so-called *roof effect* is clearly observed in the second and third spectra)

difference between the two nuclei involved is not very large compared to the *J*, the rules no longer apply, and they are considered to be *strongly coupled*. This strong coupling distorts the intensities of the simple doublets between two coupled protons in the NMR spectrum, giving rise to the situation known as the *roof effect* (Fig. 2.6, second and third spectra from the top). In this situation, the intensities of the outer resonances decrease while the inner ones increase, reaching a limit where the two coupled protons are identical (magnetically equivalent) and a single resonance is observed for both (Fig. 2.6, bottom spectrum).

2.2.2 How Does Spin–Spin Coupling Occur?

Spin–spin coupling is mediated via the electrons of covalent bonds, and therefore its magnitude decreases rapidly as the number of intervening bonds increases. The basic mechanism that propagates coupling is polarization. If there is a coupling between the electron and the nucleus, the polarization may be: (i) a dipolar interaction between the electron and the nucleus; or (ii) a *Fermi-contact* interaction (an interaction that depends on the proximity of an electron to the nucleus, and hence only observed in s-orbitals occupied by electrons). This latter interaction between electrons and nuclei does not necessarily imply that both particles must be antiparallel or parallel, as the arrangement depends on the type of atom involved in the bond. If we consider two spin-½ nuclei, A and B, linked by an electron-pair bond and presenting a scalar coupling $^1J_{AB}$, the molecular orbital between the atoms will be occupied by the two electrons of the bond, which according to Pauli's principle, will present an antiparallel spin arrangement in the orbital. The nuclear spin B can have a spin parallel or antiparallel to the spin of the second electron depending whether it is in the α or β state (Fig. 2.7a). If we assume that parallel arrangements of spins (nuclear and electrons) have a higher energy than the antiparallel ones, then two conformations are possible (and hence give rise to a doublet). The same occurs when the transmission through the bond is explained from the point of view of nucleus B. A stronger Fermi interaction will occur when the percentage of *s-orbitals* in the intervening bond is larger. For instance, for $^1J(^1\text{H}-^{13}\text{C})$, the higher the percentage of s-orbital in the hybrid orbital, the larger the spin–spin coupling constant will be, going from around 250 Hz in sp-orbitals to approximately 140 Hz in hybrids sp^3 (Günther 1995). From the above discussion, it is clear that the coupling of a nucleus to an electron via the Fermi-contact interaction is important for protons, but not necessarily relevant for other nuclei. Heavier nuclei interact through other processes like the dipolar mechanism between the electron magnetic moment and their orbital motion.

When we consider the coupling between atoms via two bonds ($^2J_{AB}$), as in H–C–H (geminal protons), we need to take into account a mechanism which "transmits" spin alignment through the central C atom. In the example shown in Fig. 2.7b, the H atom (A) polarizes the nearest electron with an antiparallel arrangement, and consequently the other electron occupying the same molecular

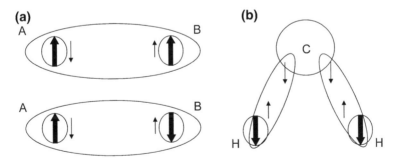

Fig. 2.7 ^1J and ^2J couplings. **a** The polarization mechanism for the spin–spin coupling ($^1J_{AB}$). The information is transmitted through the Pauli's principle, and the two possible arrangements shown have different energies. **b** The polarization mechanism for $^2J_{HH}$ spin–spin coupling. The coupling information is transmitted from one bond to the next taking into account the Hund's rule of maximum multiplicity. In both (a) and (b) panels the nuclear spins are shown by thick *black arrows* and the electrons by thin *arrows*

orbital, likely to be close to the carbon, will also have an antiparallel arrangement (Pauli's principle). The carbon atom shares another electron with the second H atom; according to Hund's rule, the electrons closer to the carbon, in both C–H bonds, must have parallel orientations. In the second C–H molecular orbital, the electrons must have an antiparallel arrangement according to Pauli's principle, and therefore, the energy of the second H nucleus (atom B) is the result of the spin arrangement with its closest electron. Of course, the value of the interaction (and hence, the magnitude of ^2J and its sign) will depend on the hybridization of the carbon atom.

2.2.3 Variations in the Value of J

We shall describe briefly how the spin–spin coupling constant varies according to chemical and structural factors. We shall focus on couplings involving protons, since the body of literature on that nucleus is comprehensive (Günther 1995; Sternhell 1969; Barfield et al. 1990; Homans 1989).

(a) *Geminal couplings* (^2J). This corresponds to the coupling between two protons bound to the same carbon atom. Values rely heavily on the hybridization of the carbon nuclei to which both H nuclei are bound and on the presence of electronegative substituents. To be able to observe the coupling between two protons, they must experience different magnetic environments. The range of ^2J values varies from 12–20 Hz in geminal protons in alkanes to 0–3.5 Hz in geminal protons in alkenes (Günther 1995).

(b) *Vicinal couplings* (^3J). The coupling between protons separated by two carbon (or other) atoms. The value of this coupling constant depends on: (i) the presence of electronegative substituents; (ii) the distance between the two carbons involved (r_{CC}); (iii) the HCC valence angles (Θ_1 and Θ_2); and (iv) the dihedral angle between the C–H bonds under consideration (ϕ) (Fig. 2.8a). Perhaps the most distinctive feature is the variation with the ϕ angle, which varies with the expression: $^3J = A + B\cos(\phi) + C\cos(2\phi)$, where A, B and C are constants that depend of the nuclei involved (Fig. 2.8b). This mathematical expression is known as the *Karplus equation* and its use requires the assumption that the dihedral angle adopts a prevailing value over time (that is, a single rotamer must be predominantly populated). If there is a rapid averaging of the dihedral angle (that is, the bonds are easily rotatable) an averaging of the possible coupling constants will be observed. The equation can be successfully applied to systems where the bonds cannot be rotated (in sugars, for instance, the vicinal coupling constant between the anomeric proton and its neighbour determines whether the sugar is α or β). The equation is explained by the fact that the overlapping efficiency of the molecular orbitals involved in "transmitting" the coupling is determined by the ϕ angle.

(c) *Long range couplings* (^nJ, $n \geq 4$). Although generally small in size, ^nJ couplings are frequently observed. They normally produce very small splittings, which are observed at high magnetic fields. Currently, splits of 0.2 Hz or even

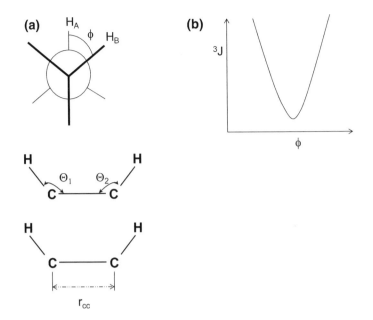

Fig. 2.8 Variation of ^3J. **a** Factors on which ^3J depends (from *top* to *bottom*): dihedral angle (ϕ), valence angles (Θ) and the carbon-carbon distance (r_{cc}). **b** The Karplus curve (the angles must be measured in radians) shows the variation of ^3J according to the dihedral angle

lower can be easily detected, and in general the ^4J values are in the range 0.1–3 Hz, although larger values are not unusual. They are commonly observed in saturated atoms which are coplanar or in a W-shaped conformation, and also in atoms involving π-orbital bonding. Although in principle π-electrons do not propagate scalar couplings because π-orbitals have nodes (no electronic density) at the position of the nuclei, ^4J can be measured in these systems due to the high delocalization of the electronic density. A further description of the mechanism of transmission in these coupling constants through the bonds is beyond the scope of this chapter but the interested reader can find further details in the literature (Günther 1995).

2.2.4 Spin–Spin Decoupling

Spin–spin coupling yields important chemical information about the number of nuclei surrounding a particular observed nucleus, but they also complicate the NMR spectra by increasing the number of lines. Therefore, techniques used for eliminating spin–spin coupling are of considerable importance, and are routinely used in 1D, 2D and nD-NMR experiments (Chaps. 3 and 4). Decoupling can be applied to the ^1H nucleus when collecting, for instance, ^{13}C spectra, but the scheme is of general applicability to any other nuclei. An in-depth description of available decoupling techniques can be found in the literature (Freeman 1997).

Spin–spin decoupling rests on the application of a second radiofrequency source, in addition to the transmitter frequency used for the recording of the main spectrum. If we consider a spin system formed by two spins A and B showing a J_{AB} coupling, a normal spectrum will present a doublet for each of the spins. If during the recording of the NMR spectrum we irradiate at the precise frequency of, for instance, the B nucleus with the second transmitter, the effect will be the disappearance of the coupling at the A frequency which will show a single line. During the irradiation at B, this nucleus is switched around by the decoupling field so fast (compared to J), that the A spins only "see" the average of the B spin, that is, the probability of an upward transition in B is the same as for a downward transition, with the result of the B multiplet being saturated (Chap. 1). An example of the decoupling technique is shown in the inset of Fig. 2.1 where the CH_2 signal of ethanol has been decoupled (the bolt). In this case, the saturation of the methylene resonance removes the splittings from the coupled signals, yielding two singlets for both the CH_3 and OH signals.

Although decoupling is a very useful NMR technique, it can also cause problems. Incomplete decoupling leads to signal broadening, imperfect selectivity of the irradiated signal leads to decoupling of neighbouring resonances, changes in the chemical shift of the signals close to the irradiation peak (Bloch-Siegert shifts) and population alterations which can vary the intensity of some resonances due to, among others, NOE effects.

2.3 The Nuclear Overhauser Effect

In the Sect. 2.2 we have described the scalar couplings between nuclei: indirect couplings are transmitted through the electrons intervening in the chemical bonds. In this section, we shall be dealing with another class of coupling: the dipolar coupling that gives rise to the *nuclear Overhauser effect* or *NOE*. The effect is described simply by considering a system formed by two spins A and X, which do not show spin–spin scalar coupling, but are spatially close enough to interact by means of a dipole-dipole interaction (Chap. 1). In the NMR spectrum of such system we should expect two lines, one from A and another from X. If we irradiate the system with a radiofrequency at the resonance frequency of X, providing that enough power is used to saturate the transition (that is, the populations of the X levels are equal, Chap. 1), we shall observe that the intensity of the A resonance is modified: enhanced, diminished or even converted into an emission line rather than an absorption one. This modification is the NOE.

The NOE measurement is crucial to the determination of the conformation of molecules. This information can be obtained because the NOE depends, among other factors, on the distance between the nuclei involved. The NOE effect is also used as a way of enhancing the sensitivity of low-γ nuclei. For instance, the decoupling of ^1H during the recording of ^{13}C spectra, contributes to an increase in the ^{13}C signals by means of the NOE effect between each carbon and its attached protons. We shall describe in this section the basis and main applications of the NOE, but for a full description of the matter, the interested reader is directed to the comprehensive text by Neuhaus and Williamson (Neuhaus and Williamson 1999) which also includes important applications of the NOE in the study of biomolecules.

For the description of the NOE background, how the effect arises and what factors determine its sign and magnitude, we shall restrict our discussion to *steady-state* NOEs. In these NOEs, the perturbation is brought about by saturating one of the spins via the selective application of a weak radiofrequency irradiation at the frequency of that particular spin resonance. This is the kind of NOE observed in the well-known method among chemists, as the NOE-difference. We shall begin the discussion by considering the simplest example of a homonuclear spin-½ system, and from there we will extend to more representative multispin systems. Further details and the kind of experiments used will be described in Chaps. 3 and 4.

2.3.1 The Basis of the NOE: A Two-Spin System

To be able to understand the NOE, we need to think in terms of energy level populations. In an AX system formed by two spins which are not scalarly coupled, there will be four energy levels (Fig. 2.9a). At equilibrium, the population of the $\alpha_A\alpha_X$ state is the largest, and that of the $\beta_A\beta_X$ state is the lowest, with the other two levels showing the same energy and an intermediate population. The intensities of the peaks in the NMR spectrum (Fig. 2.9a, bottom) reflect the population in each of the levels (Chap. 1). When the X transitions are saturated (Fig. 2.9a, bolts), the populations of the X levels become identical, but no changes are observed in the population of the A levels (Fig. 2.9b). In a system with two isolated spins, this X saturation would cause in the spectrum the loss of the X resonance and no effect whatsoever on the A resonance (Fig. 2.9b, bottom). However, in our dipolarly coupled A-X system we have to take into account the relaxation effects between the two spins (Chap. 1). Since the system has been forced away from equilibrium by altering its level populations, it will be inclined to strive to regain the original state by recovering those spin populations in each level via relaxation processes.

Relaxation can occur in several ways if dipolar interactions exist between the A and X spins. However, to be observed in an NMR experiment, the process must correspond to a single-quantum transition. This process involves a net change of quantum moment of 1, that is, it implies the flip of a single spin (A or X), as for example from $\alpha_A\alpha_X$ to $\beta_A\alpha_X$. There are two of these processes in the system (Fig. 2.9c). However, there are another two additional relaxation processes by which the system can regain the equilibrium population. For instance, another relaxation possibility is for an active field to cause both spins to flip concomitantly from β to α, and therefore, the $\alpha_A\alpha_X$ and $\beta_A\beta_X$ levels will regain their equilibrium

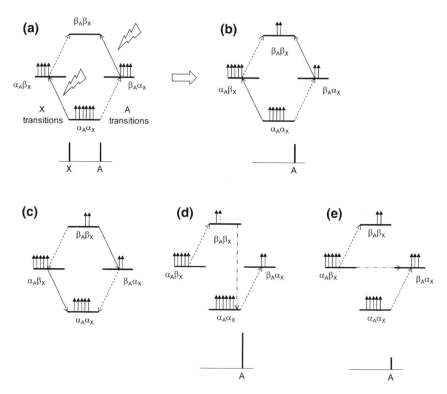

Fig. 2.9 The NOE. **a** The energy levels of an AX system and an indication of their relative populations. The transitions involving the changes in the A spin are indicated by dashed arrows while those involving the change in the X state are shown as continuous arrows. The bolts indicate the transitions being saturated (those of nucleus X). The corresponding spectrum is shown at the bottom. **b** Energy levels after saturation of the X transitions. The bottom of the figure shows the resulting spectrum. **c** Single-quantum transitions, which could induce the recovery of the equilibrium populations. The continuous or dashed arrows have the same meaning as in panel (a), but their direction is opposite (with the arrow head pointing towards the lower energy level). **d** The double-quantum transition is shown as a dot-and-dash arrow. The dotted spin in the highest level indicates the one which is transferred at the lowest level; now the transitions (dashed arrows) involving the A spin are more intense because there are more spins in the basal A levels than in the equilibrium situation (panel (a)). The bottom of the figure shows the resulting spectrum (positive NOE). **e** The zero-quantum transition is shown as a dot-and-dash arrow. The dotted spin in the medium level indicates which one is being transferred; in this case, the transitions (dashed arrows) involving the A spin are less intense (because there are fewer spins in the basal A levels) than in the equilibrium situation (panel (a)). The bottom of the figure shows the resulting spectrum (negative NOE)

populations. This is called a two-quantum transition, since the orientation of the two spins change concomitantly and two quanta are changed. This transition cannot be observed directly in an NMR experiment, as there is a net change of two in the total quantum moment (the transition is said to be forbidden by quantum mechanics rules) (Fig. 2.9d). However, this spin flipping leaves the populations of the saturated $\alpha_A\beta_X$ and $\beta_A\alpha_X$ unaltered; therefore, the population difference between the states joined by changes in the spin A are now larger than at equilibrium (there are more spins in the basal state). As a consequence, the intensity of the resonance absorption will be enhanced (Fig. 2.9d, bottom). This condition represents a NOE effect where the difference in intensity results in signal enhancement.

The last possible relaxation mechanism between the levels is the dipolar interaction causing α to flip to β and β to flip to α, or the $\alpha_A\beta_X$ - $\beta_A\alpha_X$ transition. This zero-quantum transition (the net change of quantum moment is null) is not observed directly in the NMR experiment either (Fig. 2.9e). The zero-quantum transition equilibrates the populations of $\alpha_A\beta_X$ and $\beta_A\alpha_X$ but leaves the populations of $\alpha_A\alpha_X$ and $\beta_A\beta_X$ unaltered. In this case, the population difference between the states involving changes of the spin A is smaller than at equilibrium and the intensity of the resonance absorption will be diminished: in this NOE effect, the difference in intensity results in signal decrease (Fig. 2.9e, bottom).

In practice, the effect lies somewhere between these two limits, and it is usually reported in terms of the $\eta(X)$ (being X the saturated nucleus): $\eta(X) = \left(\frac{I-I_0}{I_0}\right)$, where I_0 is the intensity of the resonance in the absence of any saturation and I is the NOE intensity of a particular transition. From the above considerations, it is clear that the building-up of a NOE (either negative or positive) will depend on the different contributions of the zero- or double- quantum processes (the so-called *cross-relaxation* processes). If the double-quantum contribution dominates, the NOE will be positive, whereas it will be negative if the zero-quantum relaxation contributes most. In addition, the single-quantum relaxation processes (those shown in Fig. 2.9c) will re-establish the equilibrium once the saturation has been removed, and therefore will compete with both cross-relaxation processes. If these single-quantum processes are efficient enough, NOEs will fail to develop.

To summarize all of the above, we can define the NOE as the result of a balance of different competing relaxation pathways.

2.3.1.1   The Dipolar Coupling and the Tumbling of the Molecule

Nuclear spins suffer relaxation processes caused by local-field oscillations at a frequency close to their Larmor frequencies (Chap. 1). These variations originate, in most cases, with the tumbling of the molecule in solution (defined by the correlation time, τ_c, Chap. 1). The tendency within a molecule to induce quantum transitions by means of its molecular tumbling is called *spectral density*, and is represented as $J(\omega)$. The spectral density can also be viewed as the probability of a component of the motion of the molecule being at a particular frequency. Therefore, relaxation can only occur when a frequency exists at the Larmor frequency of the particular spin, out of the whole set of frequencies that are explored by the molecule (and which are given by $J(\omega)$).

The spectral density function has the general form $J(\omega) = \left(\frac{2\tau_c}{1+\omega^2\tau_c^2}\right)$ (Levitt 2009), and the shape it adopts, depending on the value of τ_c, is shown in Fig. 2.10a. For molecules which tumble slowly (long τ_c, slow motion) there is a very small probability of finding frequencies of rapidly oscillating fields (that is, the $J(\omega)$ curve has a small area at the largest frequencies). For instance, for a transition occurring at a frequency ω_t (Fig. 2.10a) the probability of finding oscillating fields in molecules with slow motions is very small; the probability of finding species with the proper frequency will be larger only in molecules with intermediate motions (second curve in Fig. 2.10a). Using these arguments, it is possible to predict whether the zero- or double-quantum processes will be dominant in the NOE building-up. In the zero-quantum process, the energy differences between the two frequencies for the A and X spins are small (in fact, we have assumed above that the $\alpha_A\beta_X$ and $\beta_A\alpha_X$ were degenerate). Therefore, the particular ω_t will be on the left-hand side of the x-axis of $log(\omega)$ (since in a zero-quantum transition, the $\omega_t = \omega_A - \omega_X$) and we should be able to find a large number of fluctuating fields with that proper small frequency for molecules with slow motion. The relaxation mechanism governed by the zero-quantum transitions is predominant in molecules with long τ_c, which will give rise to negative NOEs.

Fig. 2.10 $J(\omega)$ and NOE curves. **a** Schematic $J(\omega)$ for three molecules tumbling with different motional regimes as a function of the frequency ω; ω_t represents the frequency of the transition. **b** Variation of the maximum theoretical homonuclear steady-state NOE in a two-spin system as a function of $\omega_0\tau_c$

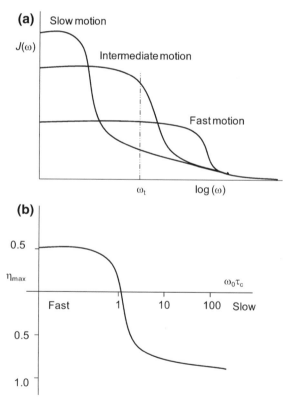

Double-quantum transitions are likely to have very large $log(\omega)$ values since ω_t will be the sum of the A and X frequencies, $\omega_t = \omega_A + \omega_X$, and will be located on the right-hand side of the x-axis of $J(\omega)$. Therefore, the probability of finding spins with the proper oscillating frequency will be higher in molecules with fast motion (short τ_c). Molecules with fast tumbling in solution will exhibit the double-quantum process as the main relaxation mechanism and give rise to positive NOEs.

Quantitative calculations show that the cross-relaxation rates not only depend on the $\omega_A \pm \omega_X$ values, but also on γ_A, γ_X and r^{-6}, where r is the distance separation between the two dipoles (A and X nuclei). This distance dependence comes from the variation of the dipolar relaxation rates (Chap. 1, Fig. 1.3), and causes the NOE to decrease sharply with distance. In practice, only NOEs between nuclei within 5–6 Å of each other will develop. The fact that cross-relaxation rates rely on the gyromagnetic constants of both spins means that very different rates can occur in heteronuclear systems. This is why decoupling of protons during the acquisition of ^{13}C spectra yields NOE enhancements of nearly 200 % ($\gamma_H \approx 4\gamma_C$), equivalent to a three-fold intensity increase of the ^{13}C signal. In heteronuclear systems, the observed NOEs will also depend on the sign of the gyromagnetic ratios involved, and in the case of fast tumbling molecules, the NOEs from protons to nuclei with $\gamma < 0$ can yield negative NOE values.

In a homonuclear system, the variation of the steady-state NOE (η) with the molecular motion, expressed by the product of $\omega_0\tau_c$ (where ω_0 is the spectrometer frequency) follows the curve shown in Fig. 2.10b. Since ω_0 is constant, the curve shows the variation with the tumbling of the molecule: if τ_c is very long (the molecule tumbles slowly), we get negative NOEs. On the other hand, if the molecule tumbles very fast (short τ_c) the NOE becomes positive, which in NMR terminology is called the *extreme narrowing limit*. In the intermediate tumbling region, the NOE changes signs and will approach zero when the double- and zero-quantum relaxation mechanisms are equal. In this intermediate region the magnitude of the NOE is highly sensitive to the solution conditions (viscosity, temperature, pH or ionic strength) as well as the shape of the molecule. In fact, for some medium-size molecules the NOE will be too weak to be observed. In these cases, it is advisable to change the experimental conditions. Changing the field strength (i.e. B_0, which results in changes of ω_0) is not always feasible; alternatively, it is possible to use NOE experiments based on the rotating-frame, which always yield positive NOEs (Neuhaus and Williamson 1999; Levitt 2009) as will be explained in Chap. 4.

2.3.2 NOEs in Multispin Systems

The above discussion on a two-spin system assumes that the main mechanism of relaxation occurs through dipole–dipole relaxation. However, the presence of other spins influences the NOE build-up and the cross-relaxation mechanisms by the intervention of other competing relaxation processes (Chap. 1). Thus, two spins showing a strong dipolar coupling due to their proximity might not produce a large NOE if other neighbours are available to compete. We shall further comment on these effects in Chap. 3, but the interested reader can also have a look at selected literature (Neuhaus and Williamson 1999; Kalk and Berendsen 1976; Cavanagh et al. 1996).

Finally, the build-up of NOEs can be affected by *chemical-exchange* processes. Exchange processes are chemical reactions that take place within the same molecule or with other chemical entities in solution. The involvement of some spins in exchange processes, and also in the building-up of the NOE, translates into saturation transfer processes. For instance, a saturation transfer can occur when a molecule suffers a conformational-exchange reaction (e.g. conformer equilibrium of sugar rings), where saturation of a spin transition in one conformer can lead to saturation in the other conformer.

References

Barfield M, Collins MJ, Gready JE, Hatton PM, Sternhell S, Tansey CW (1990) NMR studies of bond orders. Pure Appl Chem 62:463–466

Cavanagh J, Fairbrother WJ, Palmer AG 3rd, Skelton N (1996) Protein NMR spectroscopy. Principles and practice, 1st edn. Academic Press, New York

Freeman R (1997) Spin choreography: basic steps in high resolution NMR. Spektrum, Oxford

Günther H (1995) NMR spectroscopy: basic principles, concepts and applications in chemistry, 2nd edn. Wiley and Sons, New York

Homans SW (1989) A dictionary of concepts in NMR. Clarendon Press, Oxford

Kalk A, Berendsen HJC (1976) Proton magnetic relaxation and spin diffusion in proteins. J Magn Reson 24:343–366

Levitt MH (2009) Spin dynamics: basis of nuclear magnetic resonance, 2nd edn. Wiley, Chichester

Neuhaus D, Williamson MP (1999) The nuclear Overhauser effect in structural and conformational analysis, 2nd edn. VCH Publishers, New York

Sternhell S (1969) Correlation of interproton spin–spin constants with structure. Q Rev Chem Soc 23:236–270

Wishart DS, Bigam CG, Yao J, Abildgaard F, Dyson HJ, Oldfield E, Markley JL, Sykes BD (1995) ^{1}H, ^{13}C and ^{15}N chemical shift referencing in biomolecular NMR. J Biomol NMR 6:135–140

Chapter 3
Basic NMR Experiments

Abstract In this chapter we shall describe basic NMR experiments used in the fields of Chemistry, Biochemistry and Biology. The topics of acquisition and processing of simple 1D-NMR spectra will be introduced, as well as the fundamental principles behind n-dimensional NMR experiments. For 2D-NMR, some homonuclear and heteronuclear shift correlation experiments based on scalar couplings are shown. The basis of the population transfer that is at the root of heteronuclear experiments is described. The chapter ends with the most relevant nuclear correlations via dipolar couplings, based on the NOE effect.

Keywords Decoupling · Double resonance · Gyromagnetic · Heteronuclear correlation · Multidimensional experiments · Multiple quantum · NOE · Population inversion · Spectral resolution · Signal-to-noise

3.1 Introduction

In the two preceding chapters, we have introduced several characteristics of the atomic nuclei concerning their interaction with a magnetic field, such as nuclear spin, magnetization, chemical shift, spin coupling and relaxation (Chaps. 1 and 2). Although these NMR parameters are very relevant in their own right from a basic physics perspective, in this chapter we are mainly interested in the myriad of applications and experiments that put to use the measurement and interpretation of these key NMR features. The boom in useful experiments in the last decades has ensured the expansion of NMR from its Physics roots into diverse scientific fields like Chemistry, Biology, Materials and Medicine. One of the unique aspects of magnetic resonance spectroscopy compared to other spectroscopic and analytical techniques is the permanent evolution, improvement and development of applications that are specifically tailored to answer new arising questions.

The inherent flexibility of the NMR techniques allows the spectroscopist not only to employ an existing method that might provide the answers sought, but also

R. J. Carbajo and J. L. Neira, *NMR for Chemists and Biologists*,
SpringerBriefs in Biochemistry and Molecular Biology,
DOI: 10.1007/978-94-007-6976-2_3, © The Author(s) 2013

to modify an existing experiment or even *create* a new one that will address the specific problem to be tackled. As a result of this open ingenuity, thousands of different NMR experiments are now on hand that can, to begin with, seem overwhelming to the non-specialist. Several compilations of NMR experiments are available (Berger and Braun 2003; Parella 1999) where some of the most relevant methods are described in detail. Far from such comprehensiveness, our intention here is to provide a general feel for the NMR technique and its potential as analytical tool. In this chapter we shall focus on basic NMR tools applied to molecules in solution. We shall start with the acquisition of typical unidimensional (1D) spectra of some of the most common nuclei (^1H, ^{13}C, ^{19}F, ^{31}P) that provide NMR parameters such as chemical shifts and spin-couplings, followed by a description of the origin and applications of two-dimensional (2D) experiments which render more complex and, at the same time, useful information.

3.2 1D NMR

One of the visual representations of data most commonly found in daily life plots the intensity (or variation) of a parameter along an axis. Its simplicity allows an easy and intuitive interpretation of the data, and examples are plentiful: variation of stock values *versus* business hours in finances, ground motion against time during earthquakes as represented in seismographs or evolution of atmospheric pressure throughout the day in weather forecasts. These graphs can be found everywhere in science, and especially in the case of analytical techniques unidimensional (1D) representations are common. Thus, UV and IR spectra show the variation in absorbance *versus* wavelength in nm and cm^{-1}, respectively; Mass Spectrometry presents intensity *versus* mass-to-charge ratio (m/z); and in chromatographic techniques, molecular concentration (usually measured as absorbance) is plotted against retention time (or volume). In the case of NMR, the typical 1D spectrum covers the frequency range (in Hz or ppm) of the nucleus being measured and shows signals at the corresponding chemical shifts for the molecule under study, where the intensity, shape and multiplicity are related to molecular features. The position (i.e. chemical shift or frequency) of each resonance is dictated by the chemical environment of the nuclei (Chap. 2). The magnitude (intensity or integral) of each of the signals correlates with the relative abundance of the nucleus in the molecule under study, and the signal splitting (or its absence!) is a result of the chemical bonds of the nuclei.

3.2.1 Sensitivity and Frequency

Each chemical element in the periodic table has at least one isotope that is NMR active; therefore any nucleus is potentially detectable, but by no means equally accessible by magnetic resonance (Table 1.2). To compare the ease of measurement of nuclei by NMR, the concept of *receptivity* D^C was introduced:

$$D^C \propto A\gamma^3(I + 1)$$

where A and γ are the natural abundance and gyromagnetic constant of the specific isotope, respectively (Table 1.2). The reference receptivity of a nucleus is that of $^{13}C = 1$. Increase in γ value leads to greater sensitivity because of increase in dipolar moment, increased precession rate and augmented energetic transition difference.

Thus, spin $I = \frac{1}{2}$ nuclei with high natural abundances and γ (e.g. ^1H, ^{31}P, ^{19}F) will not present particular sensitivity difficulties for their detection by NMR in principle. On the contrary, direct NMR detection will be harder for nuclei with very low natural abundance such as ^{15}N ($I = 1/2$, A = 0.37 %, $\gamma = -2.71$, $D^C = 2.19 \times 10^{-2}$) or nearly impractical if both A and γ are very small as is the case of ^{187}Os, the nucleus with the lowest receptivity ($I = 1/2$, A = 1.96 %, $\gamma = 0.619$, $D^C = 1.43 \times 10^{-3}$).

From the definition of receptivity, it could be expected that nuclei with quantum spin number $I > \frac{1}{2}$ would present a significant sensitivity advantage. However, these $I > \frac{1}{2}$ nuclei (also known as *quadrupolar*) have inherent physical properties that give rise to detection problems by NMR (Chaps. 1 and 2). Their broad signals mean low precision in chemical shift and spin coupling measurements and low sensitivity. Therefore, a nucleus like ^{14}N with high receptivity ($I = 1$, A = 99.63 %, $\gamma = 1.93$, $D^C = 5.69$) suffers from a large quadrupolar moment that mars its detection by NMR. Although around 75 % of the nuclei in the periodic table have spin $I > \frac{1}{2}$, they have been much less studied by NMR than those of $I = \frac{1}{2}$.

The 1D spectrum covers a frequency range for each nucleus that depends on the γ of the isotope measured. If we imagine the whole NMR frequency range available for a 100 MHz spectrometer as a one metre line, ^1H will resonate at the end of that line (in the 100 MHz area) covering a region of around 1000 Hz, equivalent to 1 mm of the line (0.1 %). The rest of the nuclei frequencies will be distributed along that metre without overlapping, each of them spanning up to a few mm (i.e. kHz). However, due to hardware and electronic limitations, it is not possible to cover with a single radiofrequency reading or *pulse* the whole range of nuclei, the same way we cannot listen to different radio stations at the same time from a single radio device. Therefore, to acquire the spectrum of our nuclei of interest in a molecule, for example ^1H, ^{13}C or ^{31}P, we will need to record an independent 1D experiment for each of them.

The higher the spectrometer field, the longer the line representing the frequencies. However, it is important to bear in mind that, although nuclei will be located always in the same region of the imaginary frequency line relative to each other, their absolute frequencies and range will vary with the spectrometer field; that is, the frequency of the ^1H will be always located at the end of the range, but this will correspond to 1000 Hz at 100 MHz and 6000 Hz in a 600 MHz magnet.

This prompted the early NMR community to define the concept of *chemical shift* (Chap. 2). In this way, the NMR spectra of a molecule acquired under identical experimental conditions (solvent, temperature, etc.) but at different spectrometer strength will give identical chemical shift values, with the only difference in the signal resolution due to the larger number of hertz available to define the resonances at higher fields.

3.2.2 Acquisition and Processing

Before we start any NMR experiment it is necessary to tune the probe and to shim the magnet with the sample inside (Chap. 1) (Berger and Braun 2003; Derome 1987; Claridge 1999). Once lock and shims have been finely adjusted, we are ready to acquire our NMR experiments. However, there are several parameters that we will need to adjust/determine in advance to maximise signal-to-noise and resolution. In addition it is worth having post-acquisition protocols to improve the quality of the final spectrum.

3.2.2.1   Signal-to-Noise Ratio (S/N)

We have described previously how to acquire the FID of an experiment (Chap. 1). We know from other spectroscopies that the difference in population between energy levels depends on the Boltzmann constant. In the case of NMR spectroscopy, the energy difference ΔE caused by the magnetic field is very small (1 in 10^5 population difference for a 500 MHz spectrometer and ^1H), and directly responsible for the intrinsic insensitivity of the technique. Spectrometer frequency is relevant to ΔE, as increase in magnetic field increase the ΔE gap, and consequently increased population differences. The resolution of an NMR spectrum increases linearly with B_0 and the sensitivity increases with $B_0^{3/2}$, which are the main reasons for the constant technological effort towards higher field magnets. Apart from the field strength, the S/N of the spectrum depends to a large extent on the concentration of the sample, i.e. how many molecules contribute to the NMR signal, where the limiting factor is the availability and/or the solubility of the material. To overcome the low sensitivity of NMR several hardware implementations, acquisition short-cuts and processing "tricks" that improve the sensitivity of the measurements have been introduced over the years.

While other spectroscopies such as UV or IR yield the required information with single-scan acquisitions, such fast experiments are normally insufficient in NMR, and signal averaging is generally required. By accumulating a number n of scans we are adding n spectra on top of each other, with the NMR signal correspondingly increasing n times. However, electrical noise arising from different sources also contributes to the FID, but because noise is inherently random it only increases by $n^{1/2}$. Therefore, the actual gain in S/N after n accumulations is just $n^{1/2}$.

So, in order to double the S/N of a spectrum recorded with 16 scans, we will need four times that number of accumulations (64), with an associated four-fold increase in time. Amongst the several innovations introduced in modern NMR to improve the sensitivity, one of the most successful has been the use of cryogenically cooled probes. These appliances keep the receiver/transmitter coils at very low temperatures (around 70 K) while the sample remains isolated at the measurement temperature of choice, achieving a dramatic reduction of thermal noise in the circuits. Cryoprobes attain a given S/N with fewer scans than conventional ones, and even more importantly, with a concomitant time reduction.

3.2.2.2 Resolution

The quality of a spectrum in terms of its resolution (i.e. the capability of the system to distinguish different frequencies) will depend on factors such as the magnetic field, the number of points allocated to record the FID and the application of post-acquisition processing operations. In terms of resolution, magnets behave like digital cameras, the larger the field (more pixels), the better the spectrum (the picture). A picture taken twice, with four and twelve megapixel cameras, will look similar at the standard $4 \times 6''$ size, but when enlarged the 12 megapixel picture will present a better defined (less pixelated) image. Following the same analogy, the average frequency range for the ^1H nucleus spans 10 ppm independent of the field used but that range will cover 2000, 5000 and 9000 Hz in 200, 500 and 900 MHz spectrometers. Thus, at higher magnetic fields more hertz will be available to represent the same NMR signals, and resonances overlapped at low fields will be resolved with more potent spectrometers. Although this resolution improvement might not be that significant for the spectra of simple organic molecules presenting a limited number of resonances, when more complex systems such as biomolecules are studied, signal overlap is significant and the use of high field magnets is mandatory.

The digital resolution of an NMR spectrum also depends on the number of memory points allocated to represent the frequency range. Continuing with the digital camera analogy, the *quality* of a picture (i.e. its digital resolution) is determined by the allocated memory space when it is recorded. In a similar fashion we can acquire the NMR spectrum allocating different number of points, the higher the number the better resolved the frequency domain. This resolution improvement reaches a maximum that is determined by the decay of the NMR signals in the FID, and when the signals have fully decayed the allotment of extra points will just add noise to the final spectrum. Additional points mean an increased acquisition time (larger images take longer times to be stored in the memory card), and the signal may have vanished before the end of the FID recording. In everyday NMR routine, a balance is sought between resolution, memory space and acquisition time. An exception is where signals present fast decay due to short T_2^* relaxation times (Chap. 1) where the signal will have disappeared long before the end of the acquisition time. These FIDs will not have

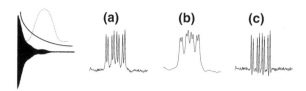

Fig. 3.1 Spectral resolution. Resulting spectra after applying apodization functions to the FID: without apodization (**a**), exponential multiplication to increase sensitivity but with resolution loss (**b**, *full line*) and Gaussian multiplication to improve resolution with a cost in sensitivity (**c**, *dotted line*)

enough frequency information to discriminate individual resonances, leading to broad signals in the spectrum (blurred pictures). Adding acquisition points will not improve the resolution in this case because the magnetization will be fully returned to equilibrium by the time the FID is recorded.

A spectrum can also be enhanced by post-acquisition manipulations of the data (Hoch and Stern 1996). Among the different "cosmetic" methods available, one of the most commonly used is *zero-filling* which consists in doubling the number of points used at acquisition by adding extra points during processing. In this way, digital resolution is increased but there is no penalty on the acquisition time of the experiment because this procedure is applied after the FID has been recorded. Although memory allocation is doubled with zero-filling, it is hardly an issue nowadays in the era of nearly unlimited disk space.

Both signal-to-noise and resolution can be enhanced by several mathematical treatments compiled by the general name of *apodization* or *window* functions. As we have previously seen (Chap. 1), maximum magnetization is observed at the beginning of the FID, which from there decays with time. Therefore, any mathematical operation aimed at increasing the preponderance of the initial points of the FID while at the same time reducing the contribution of the noise-enriched end will yield a gain in signal-to-noise, although at the cost of some loss in resolution (i.e. broader signals) as the frequency information at the end of the FID is diminished. This outcome can be achieved by the multiplication of the FID by a positive exponential function (Fig. 3.1). If the exponential function is applied with negative values, the opposite effect is accomplished: the final part of the FID is enhanced, resulting in an increase in spectral resolution (and noise) at the expense of reducing the beginning of the FID and therefore the part that contributes most to the S/N. This resolution-enhancement processing is achieved with the application of Lorentz-Gaussian multiplications. Other processing functions used for analogous purposes are available in any NMR software, such as the sinusoidal and its squared variant operation or the trapezoidal. Which functions are applied will depend on the information that is needed from a spectrum and it is not unlikely that the same spectrum will be used with different window functions representing different trade-offs.

3.2.2.3 Phasing and Integration

After the successive steps of Fourier transform, zero-filling and apodization, the spectrum is not yet ready for interpretation as it will present non-absorptive resonances. To obtain the final absorption signals, a *phasing* step has to be introduced in the process. The origins of phase errors in NMR spectra are diverse, some of them caused by the fact that the NMR signal is a complex function of real and imaginary (sine and cosine) terms (frequency dependent errors). The instrumentation also plays a part due to phase differences between the receiver and the transmitter. Likewise, due to electronic limitations, the FID cannot be collected immediately after the end of the final pulse in the NMR sequence, and a small delay has to be introduced between those two events. During that delay (commonly $10-1000$ μs), magnetization will have precessed to some extent adding extra phase errors to the final spectrum. However, semiautomatic algorithms implemented in modern NMR software packages correct these phase errors and leave the final absorptive signals. Currently, most software allows for a two-parameter phase correction that depends on the particular frequencies of signals.

As seen previously, the NMR spectrum not only contains information regarding the bonding and chemical environment in the form of chemical shift frequencies and J-couplings, but quantitative data can also be extracted from the integration of the resonances (Chap. 2). The area underneath each NMR signal is proportional to the number of atoms in the molecule that contribute to it, therefore molecular concentrations can be estimated if a reference of known concentration (e.g. TMS, DSS, TSP) is introduced in the sample. For this method to be accurate, the relaxation of nuclear spins has to be taken into account. If the succession of radiofrequency pulses is too fast, the system is not allowed to recover between scans and fully return to the equilibrium state. Therefore, for the relaxation to be complete, a pre-acquisition delay is introduced prior to the first radiofrequency pulse in order to achieve thermal equilibrium in the system. The duration of the delay (the so-called d1) will depend on the intrinsic relaxation properties of the nucleus to detect. In the case of ^{1}H, for instance, a delay five times longer than the relaxation time T_1 should provide full relaxation, and the concurrent ^{1}H spectrum can be analysed quantitatively.

3.2.3 1D Spectra of ^{1}H, ^{13}C, ^{31}P and ^{19}F

The hydrogen atom is ubiquitous in chemical space and to the advantage of NMR spectroscopists its main isotope ^{1}H presents very favourable physical properties for its detection: spin ½, nearly 100 % abundance and the highest sensitivity of the whole periodic table with a receptivity D^C nearly 6000 times that of ^{13}C. Not surprisingly, the ^{1}H signal of H_2O was the first NMR spectrum recorded, and 1D ^{1}H experiments are a major part of day-to-day routine NMR in chemistry labs. Its importance is exemplified by the fact that spectrometer strength, at least in the chemistry/biology areas, is described by the frequency at which the ^1H nucleus resonates (e.g. 500 MHz) and not by the SI unit of magnetic field, the tesla (T). A basic 1D ^1H experiment and the corresponding outcome for the cholesterol molecule are represented in Fig. 3.2a. Basically, the design of the 1D pulse sequence consists of a pre-acquisition delay (the d1 described above) that allows the establishment of equilibrium conditions between consecutive scans. This brings magnetization parallel to the magnetic field along the z axis. The d1 delay is followed by a pulse of radiofrequency applied with the carrier at the middle of the ^1H chemical shift range (Chap. 1), allowing the excitation of all proton frequencies at the same time. The effect of this pulse is to tilt the ^1H magnetization by an angle θ (commonly $90°$), with its corresponding projection on the xy-plane being detected as a FID while it decays during the time t. This pulse sequence can be repeated indefinitely and the collected FIDs added as many times as necessary to achieve the desired S/N. After the application of processing and further embellishment using the methods described in the previous section, the 1D ^1H is ready for analysis and interpretation.

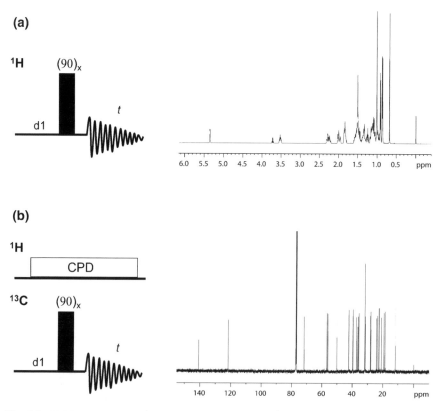

Fig. 3.2 1D NMR. **a** Standard ^1H NMR pulse sequence and ^1H spectrum of cholesterol (solvent $CDCl_3$). **b** Standard ^{13}C NMR pulse sequence and ^{13}C spectrum of cholesterol (solvent $CDCl_3$)

Carbon is also a ubiquitous element in nature, and therefore of the highest significance in chemistry and biology. Unfortunately for NMR spectroscopists, its most abundant isotope ^{12}C has null spin and therefore cannot be detected with this technique. Its measurement is possible thanks to the ^{13}C isotope, which presents spin ½ but merely 1.1 % of natural abundance. Sensitivity-wise, this number entails that only one out of every 100 molecules in the NMR tube will contribute to the ^{13}C spectrum, or if seen from a sample concentration perspective, a 1 mM sample will see its ^{1}H signals at 1 mM, but its ^{13}C atoms will be at a concentration of just 10 µM. We have described previously (Sect. 3.2.1) that the sensitivity of any nucleus depends also on its gyromagnetic ratio and in the case of ^{13}C its γ is ¼ that of ^{1}H. Therefore, disregarding natural abundance ^{13}C is ab initio four times less sensitive than ^{1}H, with a final relative sensitivity of 1.6×10^{-2} when natural abundance is taken into account. If sample availability is not an issue, good quality ^{13}C spectra can be obtained in a matter of a few hours or even less, but if there is a limitation of sample amount (µM range or below), the same task could take a prohibitive amount of time. These sensitivity issues regarding ^{13}C detection, also shared with other low sensitivity nuclei, have been partly alleviated through different improvements in hardware, such as the aforementioned cryoprobes or nano/microprobes, which use very small volumes of sample (a few microliters instead of the typical 0.5 mL) thereby making higher concentrations of sample possible where sample amounts are limiting. In the case of biomolecular NMR (Chap. 4), where the use of multidimensional and multinuclear experiments is common, these hardware implementations are not sufficient to detect insensitive nuclei such as ^{13}C (or ^{15}N), and isotope labelling of the samples is routinely practiced to achieve high levels of $^{13}C/^{15}N$ incorporation.

From an experimental point of view, a typical 1D ^{13}C experiment is acquired in a similar fashion to a ^{1}H spectrum, with the application of a short pulse at the frequency of the ^{13}C nucleus followed by the FID recording (Fig. 3.2b). A pre-acquisition delay d1 is included as well, normally of longer duration than for ^{1}H experiment, as the ^{13}C nuclei present much longer spin-lattice relaxation rates and therefore need extended times to achieve the full return to equilibrium. The ^{13}C is normally acquired with the application of ^{1}H decoupling throughout the whole pulse sequence. This broadband decoupling is accomplished via composite pulse decoupling (CPD, Sect. 3.5.3) with two main purposes: first, a sensitivity enhancement through the elimination of the splittings due to the ^{1}H-^{13}C couplings, yielding singlets for all the ^{13}C resonances independently of the number of hydrogens bound to the carbon; and second, the decoupling produces a NOE effect (Sect. 2.3.1) on the carbon from the attached Hs that also translates into an enhancement of the ^{13}C signal. For this reason, quaternary carbons present very low sensitivity in the ^{13}C spectra, as they do not receive this NOE effect, together with their longer relaxation times due to the absence of directly bound protons (and so, the absence of nearby dipole–dipole interactions). As an example, the ^{13}C spectrum of cholesterol is shown in Fig. 3.2b. The frequency range for the ^{13}C nucleus spans for over 200 ppm, and the overlap of signals is therefore less prevalent than in ^{1}H spectra (Chap. 2, Fig. 2.3b).

Analogous to the ^{1}H nucleus, chemical and bonding information can be readily extracted from the chemical shift of the ^{13}C signals applying the rule that correlates higher frequencies with increased electronegativity. Regarding coupling information, the low natural abundance of ^{13}C implies that ^{13}C-^{13}C splittings will not be observed due to the extremely low probability of finding a ^{13}C bound to another carbon-13 neighbour in all the molecules in the sample. Although the ^{1}H-^{13}C coupling is normally suppressed for sensitivity reasons (see above), the multiplicity information can be recovered by turning off the ^{1}H decoupling during the FID recording, in an experiment known as *gated decoupling*. However, this experiment suffers from very low sensitivity as the only remaining S/N enhancement on the ^{13}C signals will arise from the NOE effect produced by the ^{1}H decoupling during the d1+pulse time. Nevertheless, equivalent multiplicity information can be obtained in a more rapid and intuitive way using ^{13}C-edited experiments such as DEPT (Sect. 3.5.1). A complementary version of the gated decoupling experiment (decoupling *off* during d1+pulse, *on* during *t*) is used when quantitative measurements of ^{13}C are required. With this experimental setup the NOE contribution due to the protons bound to the carbon is eliminated and signal intensity will be exclusively ascribable to the ^{13}C natural ratio, bearing in mind that for quantitative results much longer relaxation delays will be required to achieve full recovery of the ^{13}C magnetization.

Although any element of the periodic table can theoretically be measured via 1D NMR experiments, only those with relevant interest in chemistry and/or biology that present high natural abundances and sensitivities are normally measured. Among these nuclei are ^{31}P and ^{19}F, both of spin ½ and 100 % natural abundance (Table 1.2). Their high sensitivity allows the acquisition of 1D spectra in just a few minutes, covering ranges of several hundred ppms for each of the two nuclei. The interpretation of the chemical shift variation follows the rules described for ^{1}H and ^{13}C, although in the case of ^{31}P the oxidation state of the nucleus has a notable influence that has to be considered for comprehensive interpretation. Experimentally, both ^{19}F and ^{31}P can be acquired with or without ^{1}H decoupling, but as there are no sensitivity issues regarding their detection, a non-decoupled spectrum provides extra information such as the ^{19}F/^{31}P-^{1}H *J* couplings that can be very valuable for the structural analysis of the molecule under study. ^{31}P presents an interesting characteristic related to its relaxation which is dominated by the chemical shift anisotropy mechanism (CSA) (Chap. 1). The CSA mechanism is field-dependent, and in those nuclei where it prevails the relaxation will be faster at higher fields, which can lead to broader signals. Thus, opposite to what logic would tell us, in the case of the ^{31}P detection the experiments will see an improvement both in S/N and precision with increasing spectrometer frequency but up to a certain magnetic field (around 500 MHz) above which the quality of the ^{31}P spectra declines due to the CSA relaxation dominance. Thence, higher magnetic fields (and higher costs!) are not always the best recipe for experimental success.

3.3 Multidimensional NMR

From the distance, the skyline profile of a big city like New York is largely dominated by its characteristic huge skyscrapers. That view makes the impression of the buildings being aligned side by side to each other. However, those buildings are actually located at different distances from the Manhattan shores and it is difficult to determine which buildings are closer to the observer with the only information provided by the skyline profile. In addition small buildings situated behind large ones are not visible in the profile view. Everyone would therefore agree that a skyline picture is not the best option to navigate through the streets of any city, and that a map would be a much better option. Returning to NMR, the 1D NMR spectrum is our skyline picture of a molecule, where instead of buildings we have resonances, and in place of connecting streets we have chemical shifts and *J*-couplings linking atoms. When molecules increase in size and structural complexity, more resonances will appear in the 1D spectrum and as the chemical shift range for ^{1}H is limited, some overlap will be unavoidable, with smaller signals disappearing underneath more intense ones, that hindering direct analysis of chemical shifts or couplings. In NMR, the *maps* to navigate through the spectra are multidimensional experiments, as the extra dimensions help to unravel the signal overlap and highlight connectivities. In these *n*D experiments, there is a main horizontal dimension that normally is occupied by the ^{1}H frequency range, while the extra added dimensions are constructed from the scalar and dipolar coupling connections between protons and any other nuclei within the molecule. In this section of the chapter, the basic principles to produce the second or any additional dimension in NMR will be introduced, as well as a description of some of the simplest multidimensional experiments, which are the building blocks of all the n-dimensional ones. In the following sections we will discuss some of the most popular 2D experiments, describing the kind of information they yield without going into much detail of the physics, differentiating those based in chemical bond connections such as COSY, TOCSY or HSQC from those based on dipolar couplings like NOESY.

3.3.1 Generating Dimensions in NMR

Every multidimensional NMR experiment follows the simple block scheme depicted in Fig. 3.3, with four differentiated stages: *preparation-evolution-mixing-detection*. The *preparation* stage is common to the 1D experiments described before, and includes any pre-pulse relaxation delay for the system to return to equilibrium between scans and the initial radiofrequency pulse or pulses applied to start tilting the magnetization. It is during the *evolution* stage where the multidimensional experiment is mainly fabricated. Although radiofrequency pulses can also be included during this stage, the most relevant parameter during the

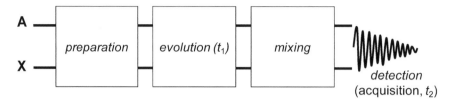

Fig. 3.3 Multidimensional NMR. Representation of a generic multidimensional NMR experiment correlating nuclei A and X. Any n-dimensional NMR experiment can be divided into four stages: *preparation, evolution, mixing* and *acquisition*. Several pulses and delays can be included into each of the stages. The convention is to represent the evolution time that gets incremented during the multidimensional experiment by t_1, whereas t_2 is the fixed acquisition time

evolution part is the incremental time t_1. The particularity of this time versus other fixed delays such as d1 is that t_1 "evolves" during the experiment, being incremented as the NMR experiment progresses. The first FID will be recorded with the shortest t_1 in the same way as any other 1D spectrum, but the following FIDs will have a total *evolution time* of $t_1 = t + nd$, where t is the starting evolution time, d is a fixed short delay and n goes from 1 to the number of FIDs we want to record to create the second dimension (number of FIDs = number of points defining the 2nd dimension). Therefore the first and last FID of a 2D experiment will have evolution times of very different lengths (we shall return later to the experimental consequences of the inclusion of an incremental delay during the sequence). The third stage within our multidimensional experiment is called the *mixing*, and during this section of the experiment that can include several pulses and short and/ or long fixed delays, the magnetization is modulated in such a way that it will render observable and interpretable information in the resulting spectrum. The last stage consists on the *detection* of the NMR signal generated along the previous steps, following the same methodology described for the 1D experiment: recording of the FID during a fixed delay (t_2) with or without simultaneous decoupling of any of the nuclei participating in the pulse sequence.

We shall now try to explain the involvement of the incremental time t_1 in the generation of the second dimension in a multidimensional experiment simply. The initial preparation stage is equivalent every time the magnetization goes through it; therefore the state of the spins at the beginning of the evolution period will be always the same. At this point we can think of the magnetization as being "labelled" with chemical shift and coupling "tags", but not yet ready to be "read". The first time magnetization goes through the evolution period, these tags will evolve during t_1 in a certain way, reaching the mixing period where they will be further transformed into a "readable" mode (the NMR signal) that is detected during the FID recording. If the Fourier Transform (FT) is applied at this point, we would obtain a 1D spectrum that corresponds with the first dimension of our multidimensional experiment. The second time we go through the pulse sequence, we will arrive at the evolution period with the same magnetization labelled with the same tags we got the first time around. However, during the second pass through the evolution period t_1 will be slightly incremented (the size of that increment is not relevant for our discussion), and the tags will have some extra time to evolve resulting in magnetization that enters the mixing period in a different state to the first time. The FT of this second FID will yield another 1D spectrum, but with incremental differences to the one recorded before. This process will be repeated n times with t_1 getting incremented nd with every step, resulting in n different FIDs. At the end of the 2D acquisition we shall have n FIDs recorded with N points that if processed will generate n 1D spectra. The second dimension is produced by selecting the first of the N points for each of the n FIDs and applying to them a second Fourier Transform, then the second point, the third, etc., till we have applied as many FTs as N points we have used to define the first dimension. The number of FIDs recorded, n, will determine the number of points we can use to define this second dimension. We define the real transformed dimension as f_2 (FID recorded during t_2) and that obtained by the increment of t_1, the f_1 dimension.

3.3.2 2D Data Acquisition and Processing

The basic tools already described in the acquisition and processing of unidimensional experiments (Sect. 3.2.2) are also applicable for the multidimensional ones. However, some aspects regarding the resolution of multiple dimensions have to be taken into account. A 2D experiment (n FIDs) will have a longer duration than the normal 1D spectrum (one FID), if keeping the number of scans constant. However, we can save acquisition time by reducing the resolution in both dimensions, which is not as crucial as in 1D experiments. For a 2D experiment a typical number of points would be 2048 (2K) for the first dimension (f_2) and 128 points for the second one (f_1), which would be equivalent to the recording of 128 1D spectra of 2K points each. The level of sensitivity needed to observe signals is defined by the number of scans applied to each FID, and considering that each acquisition normally takes a few minutes, the total 2D experimental time will be the result of those few minutes $\times$ 128. If higher resolution is needed in the second dimension (e.g. 256/512 points), the total experimental time will increase accordingly. In order to obtain higher dimensionality spectra (3D, 4D, and so on), additional incremental times are introduced in the pulse sequences, with each extra dimension needing a number of defining n points with the consequent increase in acquisition time. So while a typical 1D spectrum can last just a few minutes and a simple 2D can easily consume a few hours, 3D experiments need acquisition times in the order of days. Although improvements in hardware sensitivity or shortcuts in the acquisition/processing area have been implemented recently, it is evident from the above that multidimensional NMR spectroscopy is still a very demanding technique time-wise. Also, any extra dimension brings about an increase in the size of the recorded data that translates into increased memory allocation for each experiment (about ½ Gb for a typical 3D experiment).

We can distinguish homonuclear (same nuclei in all dimensions, normally ^{1}H) and heteronuclear (different nuclei in different dimensions) experiments. In the case of the ^{1}H homonuclear experiments, the irradiation frequency is situated in the middle of the chemical shift range of the proton, and matched spectral widths (SW) are used for all dimensions. When observing different nuclei, adequate tuning of all frequencies affected has to be done in advance, and the SW and irradiation frequency of the heteronuclei established. The latter is not a superfluous task, as the large chemical shift ranges that some nuclei require cannot be completely covered with a single radiofrequency pulse. In these cases several 2D experiments are recorded shifting in each one the irradiation frequency of the heteronucleus, covering the whole range of possible chemical shifts in this way.

As we have explained before, the resolution of a multidimensional spectrum is determined by the number of points used to define each of the participating dimensions. However, this resolution is not written in stone and can be improved a posteriori using similar tricks to those described for 1D NMR. Thus, zero-filling is normally applied in all dimensions to improve spectral resolution without a penalty in acquisition time. Another interesting mathematical operation often applied in multidimensional NMR is *linear prediction* (mathematical operation where future values of a discrete-time signal are estimated as a linear function of previous samples). In a nutshell, this operation extrapolates points at the beginning (backward) or at the end (forward) of the FID based on the actual recorded data. Therefore it adds unrecorded points to the FID, i.e. resolution, without any acquisition-time cost. However, linear prediction must be applied carefully and overuse should be avoided, bearing in mind that resolution enhancement needs a good starting S/N or it may easily lead to the generation of spurious signals.

3.4 Homonuclear Shift Correlation: Correlations Through the Chemical Bond

As the name indicates, homonuclear experiments are constructed using the same nuclear frequency, usually the ^{1}H, in all dimensions. In principle, any other nucleus could be subjected to this kind of experiment, but either their low sensitivity (e.g. ^{13}C) or the reduced number of atoms within the molecule (e.g. ^{31}P, ^{19}F) makes them much less useful than the ^{1}H variants. In the following section we describe homonuclear correlations such as COSY and TOCSY that make use of J_{HH} scalar couplings to connect protons within the molecule, and secondly briefly outline homonuclear methods such as INADEQUATE that are applied for the correlation of low-abundance nuclei.

3.4.1 COSY. Experiment Interpretation and Practical Aspects

The COSY (COrrelation SpectroscopY) sequence was the first multidimensional NMR experiment described. It was proposed by Jeener at a conference in 1971 (Jeener 1971), although its practical implementation took a few more years due to technical issues (Aue et al. 1976; Bax and Freeman 1981). Figure 3.4a shows the pulse sequence for the simplest COSY experiment and a representation of the kind of signal correlation we expect from this experiment. The basic COSY sequence presents only two $90°$ radiofrequency pulses separated by the incremental time t_1. With the first pulse (*preparation* stage) magnetization will turn from the equilibrium state in the z-axis to the xy-plane, giving rise to the *evolution* phase where spins develop chemical shifts and homonuclear coupling connections during t_1. The second $90°$ pulse (*mixing* block) converts this information into magnetization that can be analysed. The diagonal signal correlates each proton (or any other nucleus used in the COSY) with itself and therefore provides no structural or bonding information. The interesting correlations are those located off-diagonal (the cross peaks) that connect any protons that are scalarly coupled, meaning that they share a J value. These connections inform about nuclei separated by a few chemical bonds, with the intensity of the cross peak correlating with the size of the J coupling $(^2J > {}^3J > {}^4J)$. Both sides of the diagonal provide equivalent information as both dimensions share the same frequency range.

Several variations of the COSY experiment have been developed over the years. One of the most popular is the COSY-45, which receives this name due to the substitution of the second ninety degree pulse by a $45°$ one. The reduction of this pulse length achieves a simplification of the appearance of the spectrum by reducing the autocorrelation peaks on the diagonal, allowing for observation and interpretation of cross peaks located very close to the diagonal, generated by scalarly coupled protons with small chemical shift differences. Additionally, cross peaks in the COSY-45 appear tilted and the positive/negative sign of the slope indicates the number of bonds separating the connected protons, which allows differentiation between ^2J and ^3J couplings. Other COSY variant is the DQF-COSY (Fig. 3.4b) where an extra pulse is added in the mixing step creating *double-quantum* transitions or coherences (Rance et al. 1983) (see Chap. 2 for a brief description of *double-quantum* transitions in the relaxation pathways of NOESY). Although a detailed description of these types of transitions is out of the scope of this book, it is interesting to know that the water molecule has no double-quantum transitions, which allows it to be eliminated from the DQF-COSY spectrum, while the rest of cross peaks are retained. Therefore, this sequence is specially recommended for samples dissolved in water, such as those used in biomolecular NMR. Furthermore, the DQF diagonal shows a significant reduction of the characteristic *tails* in the cross peaks that are common in the typical COSY experiment. The fine structure of the cross peaks allows for another level of interpretation.

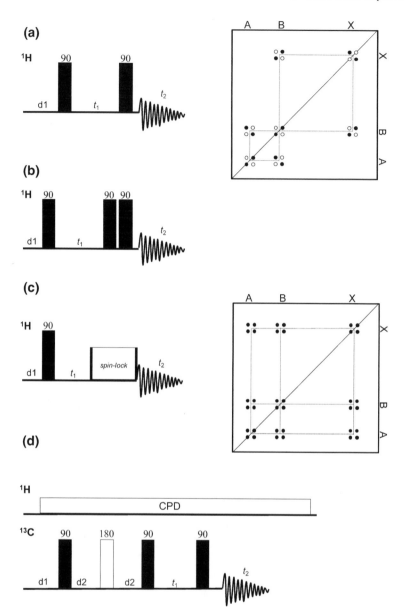

Fig. 3.4 2D homonuclear shift-correlated experiments. **a** Basic 2D ^{1}H-^{1}H COSY pulse sequence and scheme of the expected spectrum for an A-B-X spin system, where A-B and B-X are scalarly coupled but A-X are not. The pattern of the peaks in a COSY corresponds to four lobules, with alternated signs (positive and negative), due to the effect of the coupling constant among the covalently-bound nuclei. **b** DQF-COSY pulse sequence. **c** ^{1}H-^{1}H TOCSY and pattern of peaks for the A-B-X spin system. The spin-lock is achieved via a composite pulse sequence consisting in a train of several consecutive low power pulses; the pattern of peaks corresponds to a single lobule with the same sign. **d** 2D ^{13}C-^{13}C INADEQUATE sequence; it operates via the creation of multiple quantum coherences (MQC); the delay d2 is optimized for a value $1/4J_{CC}$

3.4.2 TOCSY. Practical Aspects

The TOCSY pulse sequence (TOtal Correlation SpectroscopY) (Fig. 3.4c), also known as HOHAHA (HOmonuclear HArtmann-HAhn), is a homonuclear experiment that provides a complete connection of all the protons participating in a particular spin system (Braunschweiler and Ernst 1983). The correlation is produced via scalar couplings, but in contrast to the COSY experiment which yields cross peaks only when a direct *J* coupling exists between nuclei, in TOCSY the correlation between spins can be observed even in the absence of a direct coupling, as long as a third spin is coupled to both. The preparation and evolution sections of the pulse sequence are identical to the COSY scheme, but they differ in the mixing stage. During the mixing period a composite pulse sequence (MLEV or DIPSI) is applied, consisting of a train of many consecutive low power pulses or *spin-lock*. This scheme makes the spins only *sense* B_1 (created by the train of pulses) as the effective field, eliminating any chemical shift differences between connected protons. During this time, the spins are *locked*, and cross peaks are produced with spins within the whole spin system. Experimentally, the duration of the spin-lock period determines how far into the spin-system the protons mix with each other, but this parameter also depends on the size of the molecule under study. Spin-lock blocks ranging between 40 and 200 ms are typical, with the shorter ones mainly used for large biomolecules and the longer ones applied to small molecules. However, it is good practice to acquire two TOCSYs with different duration in the spin-locking to account for relaxation effects in biomolecules (Cavanagh et al. 1996).

The TOCSY experiment gives more information than the COSY, as the scalar coupling information is relayed along the spin system. However, there is no way to discriminate which TOCSY peaks arise from direct coupling and which from relayed coupling. Therefore, combining the information rendered by both experiments gives the best results for the structural analysis of the molecules.

3.4.3 Correlation for Diluted Spins: The INADEQUATE Experiment. Double-Quantum Selection

Typical homonuclear experiments are not normally applied for low abundance spins such as ^{13}C because the probability of finding two carbon-13 nuclei scalarly coupled is very low and detecting their signal would require prohibitively long acquisition times. However, there are experiments that facilitate the correlation of dilute spins; among them is the INADEQUATE sequence (Incredible Natural Abundance DoublE QUAntum Transfer Experiment) (Fig. 3.4d). INADEQUATE achieves its target by creating multiple quantum coherences (MQC) between scalarly coupled spins (Freeman 1997). Multiple quantum effects can be fully grasped with quantum mechanics, however for illustrative purposes, some authors describe the MQC (Claridge 1999) within the naïve framework provided by the vectorial model, as antiphase vectors (see the INEPT description below) which have zero net magnetization and therefore cannot be observed (as happens with the *double-quantum* sequences). Only spin-coupled systems are capable of creating MQC, in contrast to other intense signals produced by the pulse sequence which can then be filtered out. MQCs are built during the preparation stage of the pulse sequence by the application of three radiofrequency pulses separated by delays determined by J (d2), the coupling constant connecting the spins to be detected. The preparation block is followed by the evolution one and its incremental time t_1, ending with the mixing stage where a final pulse transforms the MQCs, which are not directly detectable in the NMR, into magnetization that we can observe and interpret. The cross peaks appearing in an INADEQUATE spectrum correlate with the chemical shifts of the nuclei in the horizontal (f_2) dimension, whereas the vertical (f_1) dimension represents the sum of frequencies of two spins that are coupled (Bax et al. 1980, 1981a, b; Buddrus and Bauer 1987) Although the INADEQUATE experiment is still very insensitive and can take several days of acquisition time even for very concentrated samples, it renders very valuable information for the structural characterization of molecules. It is mainly used for systems where the application of heteronuclear correlations is not possible due to the absence of sensitive nuclei such as ^{1}H or ^{31}P in the molecule. It has found its niche in the detection of ^{13}C-^{13}C connections although other low abundance nuclei can be used, and it has also found wide applications in the field of metal NMR (e.g. ^{183}W) where it is common to find molecules lacking other sensitive sources of magnetization.

3.5 Heteronuclear Shift Correlation: Correlations Through the Chemical Bond

Chemical shift values inform about the chemical environment of nuclei, and are determined by chemical bonding and spatial interactions. Together with the information extracted from spin–spin couplings, chemical structures can be established via NMR spectroscopy. However, extracting the relevant information from crowded spectra or ambiguous data can give rise to multiple valid structural possibilities, and determination of the correct one can be a far from trivial task. We have seen previously in the description of 2D homonuclear experiments like COSY or TOCSY that the interpretation of cross peaks in those spectra is much easier than in overcrowded 1D experiments, where the ambiguity of degenerate couplings and the overlap of signals can be a dead-end for molecular assignment (i.e. the atom-NMR signal equivalence). Analogous to what we have described for homonuclear experiments, heteronuclear spectra can connect different nuclei within the same experimental setup. The ^{1}H spin normally acts as the source of magnetization in these heteronuclear experiments because of its high sensitivity and omnipresence in chemical entities. Any heteronuclei can in principle be detected using these schemes, providing a scalar coupling exists between the heteronucleus of interest and ^{1}H. However, the most common heteronuclear experiments are applied to the detection of the ^{13}C and/or ^{15}N spins (in biomolecular NMR, Chap. 4), although examples are available for most of the nuclei in the periodic table. A selection of some of the most relevant heteronuclear sequences follows. The interested reader can find more sequences in specialised texts (Muller 1979; Lerner and Bax 1986; Bax and Subramanian 1986).

3.5.1 Polarization Transfer Experiments: SPT and INEPT Sequences; Indirect Spectroscopy

The basis of all *polarization transfer* experiments is the transmission of magnetization between nuclei that are coupled. Magnetization can be transferred from a high sensitivity nucleus (e.g. ^{1}H) onto a less receptive one (e.g. ^{13}C, ^{15}N) with the aim of indirectly detecting the latter (indirect spectroscopy); the advantages of the scheme are apparent. Polarization transfer not only achieves the detection of low abundance/sensitivity nuclei but at the same time it provides chemical bonding information crucial for the assignment/identification of complex molecules. The opposite strategy (transfer from low to high sensitivity nucleus) will generate equivalent information about the system but will present evident sensitivity problems. Indirect (or inverse) spectroscopy has led to the development of most of the multidimensional experiments that are applied nowadays to medium size molecules and large macromolecules like proteins or nucleic acids. The basis of these NMR experiments is *population transfer*, which can be described as an inversion of populations of states (α and β are swapped) while observing the response from a coupled X nucleus (Fig. 3.5a). For this transfer of populations to work, a J coupling between the two nuclei is required. Although they are not currently applied as commonly as when they were first developed, the simplest population transfer schemes will be described as they constitute the building blocks from which modern pulse sequences are constructed.

3.5.1.1 SPT

In the SPT or Selective Population Transfer experiment one line in a multiplet is irradiated, altering the intensities of the lines of a second multiplet to which it is coupled (Sørensen et al. 1974; Jakobsen et al. 1974). Experimentally, this irradiation procedure is achieved via a *soft* (low power) long pulse (Fig. 3.5b). By selectively irradiating the ^{13}C satellites in the ^{1}H spectrum (arising from the ^{1}H-^{13}C coupling) and subsequently recording the ^{13}C spectrum, we shall get a net intensity gain in the ^{13}C signals. The soft pulse applied at a specific satellite resonance in the ^{1}H spectrum produces the transfer of populations, whereas the

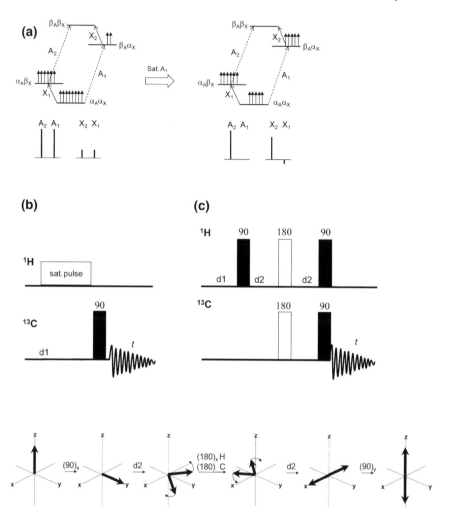

Fig. 3.5 2D heteronuclear-shift correlated experiments. **a** Energy levels, populations and expected signals for a heteronuclear AX system (e.g. ¹H-¹³C) at equilibrium (*left*) and after saturation of the A1 transition (*right*). **b** ¹H-¹³C Selective Population Transfer (SPT) pulse sequence. **c** ¹H-¹³C INEPT pulse sequence (*top*) and vector diagram for the evolution of the proton magnetization (*bottom*). After the tilting of the proton magnetization by the first 90° pulse and its evolution during d2 (1/4J_{CH}) the two 180° pulses on both nuclei flip the vectors about the x-axis and invert their sense of precession. Further evolution during the second d2 leads to antiphase vectors, which are aligned along the ±z-axis by the 90° pulse on the proton. A sensitivity enhancement will be observed in the ¹³C signals via polarization transfer

90° degree pulse on ¹³C tilts the magnetization of the carbon nucleus that is finally observed. The ¹³C spectrum recorded in this way will show those carbon signals coupled to the irradiated protons with increased intensities and modified multi-plicities compared to a normal 1D ¹³C spectrum, thereby allowing the identification of those carbon atoms that are bound to the irradiated protons. By initially exciting the more sensitive of the two nuclei in the general SPT, we obtain a gain in S/N according to the formula γ_H/γ_X that in the ^{1}H-^{13}C case reaches an enhancement factor of 4 compared to the ^{13}C direct detection experiment. The SPT is of general applicability, and for nuclei with even lower γ values than ^{13}C the gain increase is larger. In the case of molecules that do not contain protons or these are not coupled to the heteronucleus of interest, the same procedure can be applied substituting the ^{1}H for other nucleus that serves as polarization source, preferably of high sensitivity (^{31}P, ^{19}F, etc.).

3.5.1.2 INEPT

The INEPT sequence (Insensitive Nuclei Enhanced by Polarization Transfer) (Fig. 3.5c) results in a significant improvement over the simpler SPT and its development in the late seventies in the group of Ray Freeman opened the way for the subsequent development of the multidimensional heteronuclear experiments (Morris and Freeman 1979). INEPT does not employ selective polarization on individual signals as is applied in SPT, but instead it uses hard radiofrequency pulses with the aim of exciting the whole frequency range of the nucleus and achieving all possible population transfers in one single experiment. In this way the selection and precise irradiation of signals is avoided. As before, the existence of a J coupling between nuclei is needed for the experiment to produce its outcome, which is to observe the signals from the heteronuclei with a significant sensitivity enhancement by the transfer of populations from a more sensitive nucleus, normally the proton.

Although the number of pulses is increased in comparison with SPT, the INEPT maintains a very simple scheme. The first 90° pulse is applied on the more sensitive nucleus, creating transverse magnetization in the xy-plane. Two delays optimized to a length d2 $= 1/4\ J$, J being the scalar coupling between ^{1}H and X, are located at both sides of the 180° pulses applied simultaneously on both nuclei. During these two d2 delays both the ^{1}H-X spin coupling and the chemical shift evolve, but the latter is refocused by the 180° applied on the proton frequency, whereas the 180° on X modulates the spin-echo (similar to those described in Chap. 1 to measure the T_2). At the end of the second d2 delay the sequence has developed antiphase magnetization of the proton with respect to the coupled carbon, which is converted into ^{13}C antiphase magnetization by the two final 90° pulses, finally evolving into in-phase ^{13}C magnetization (detectable signal) during the FID. In short, with the INEPT scheme we shall achieve in the first place transfer of magnetization from the sensitive (^{1}H) to the insensitive nucleus (X) resulting in a net gain of S/N, and secondly we shall obtain X resonances with splittings arising from J couplings to the protons, their sizes and multiplicities revealing relevant structural information. No decoupling scheme is applied in the INEPT experiment to preserve the couplings. The increase in sensitivity is equivalent to that obtained with the SPT experiment (γ_H/γ_X) and independent of the sign of gamma.

Although the basic INEPT experiment is not usually applied on its own nowadays, its five-pulse building block is extensively used in more complicated multidimensional experiments where heteronuclear population transfer is required (HSQC and related experiments; see Sect. 3.5.2). Some INEPT variants have been developed, among them the reverse-INEPT (Freeman et al. 1981), where the population transfer goes from X to ^{1}H. Although applying the starting pulse on the insensitive X nucleus may seem counterproductive, it has the advantage that the final detection, and therefore the main source of sensitivity, is on the ^{1}H. The final spectrum will be a 1D ^{1}H, showing antiphase proton signals split according to the couplings with the heteronucleus. Hardly used in routine NMR on its own, the reverse-INEPT block of pulses is frequently used as a method to return magnetization from X to ^{1}H in more complex pulse sequences.

3.5.1.3 DEPT

The DEPT (Distortionless Enhancement by Polarization Transfer) is another experiment based on polarization transfer (Ernst et al. 1987; Doddrell et al. 1982) that is generally applied for spectral editing (selection or differentiation of NMR signals according to their chemical nature), as it provides a sign differentiation of the heteronuclei resonances according to the number of attached protons; in general terms, it can be considered an improved version of the INEPT experiment. It follows the typical scheme based on the magnetization transfer from the protons to the X heteronucleus, but it relies now on a selected flip angle of one of the pulses which will determine the sign of the X(H)n resonances. Largely, it has been applied for the differentiation of CH, CH_2, CH_3 and quaternary carbons (absent in the DEPT) in ^{13}C spectra. The most used variant is DEPT-135 (flip angle $= 135°$) that provides positive signals for the CH and CH_3 carbons and negative for the CH_2. With a straightforward superimposition with the typical 1D ^{13}C spectrum from the same molecule, a quick identification of the carbon multiplicities can be easily carried out.

3.5.2 Heteronuclear Single-Bond Correlations: HSQC and HMQC

The HSQC (Heteronuclear Single Quantum Coherence) (Bodenhausen and Ruben 1980) and HMQC (Heteronuclear Multiple Quantum Coherence) (Bax et al. 1983, 1990; Norwood et al. 1990) pulse sequences are two of the most applied experiments in routine NMR. Although obtained through different pathways, the information they provide is equivalent: a single-bond correlation between a sensitive nucleus, most commonly ^1H, with the heteronucleus of interest (e.g. ^{13}C, ^{15}N). For example, the ^1H-^{13}C heteronuclear correlation combines in a single experiment the chemical shift information contained in the corresponding 1D ^1H and ^{13}C spectra, plus the connection between the protons and the carbon to which they are bound. Indirectly, the correlation allows the identification of those protons in the molecule not bound to a carbon atom (e.g. NH, OH) and carbons with no attached proton (quaternary carbons). Together with the homonuclear experiments described before, these heteronuclear correlations are an essential tool in the structural analysis of any molecule, independent of their size and complexity.

The basic 2D HMQC experiment (Fig. 3.6a) shows a remarkable simplicity with only four radiofrequency pulses. The preparation part of the sequence includes a 90° pulse on each of the frequency channels (^1H and X), separated by a delay d2 optimized to $1/2\ J$, where J is the average one-bond ^1H-X scalar coupling. This building block achieves the creation of the multiple quantum coherences (MQC) between any two coupled spins that give name to the sequence. In the middle of the evolution time t_1 there is a 180° pulse on the ^1H channel that removes multiple quantum frequencies corresponding to ^1H chemical shifts, leaving undisturbed any ^1H-^1H coupling as well as the ^1H-X MQCs. The final 90° pulse on the X channel returns the ^1H-X MQ coherences to magnetization we can actually observe, with the FID being acquired at the ^1H frequency contrary to X-detection of the polarization transfer sequences (SPT, INEPT, DEPT) described before (Fig. 3.5). The application of the initial pulse on the sensitive ^1H nucleus and its detection on the same channel at the end of the sequence increases dramatically the sensitivity of the HMQC compared to the INEPT-type sequences. Although the HMQC generates one-bond correlations with high S/N, the detected signals are in antiphase and split due to the homonuclear ^1H-^1H couplings evolving during t_1. To obtain fully absorptive signals, the HMQC is normally processed in magnitude mode (i.e. absolute value), removing any sign discrimination but also significantly broadening the resonance peaks.

Although the HMQC experiment was originally designed for the correlation of one-bond heteronuclear couplings, the delay d2 can be adjusted to detect any long-range ^1H-X ^nJ. However, the evolved ^1J couplings will also appear on the 2D spectrum adding complexity to the molecular characterization. To be able to discriminate long-range correlations from one-bond ones, a variant of the HMQC called HMBC (Heteronuclear Multiple Bond Correlation) (Fig. 3.6b) (Bax and Summers 1986; Summers et al. 1986) has been developed. By adding a few extra pulses and delays but maintaining the simplicity of its HMQC source, the HMBC achieves the generation of heteronuclear correlations up to several bonds as far as there is a measurable scalar coupling. The two additional pulses at the beginning of the sequence act as a *low-pass* filter that is responsible for the elimination of the potentially disturbing ^1J correlations. Combined with the results from the simple ^1J HMQC, the connecting information obtained in the HMBC normally ensures the full structural assignment of small to medium sized molecules. As in the HMQC sequence, antiphase magnetization originated from the heteronuclear couplings enters the acquisition time; no refocusing or decoupling are included in

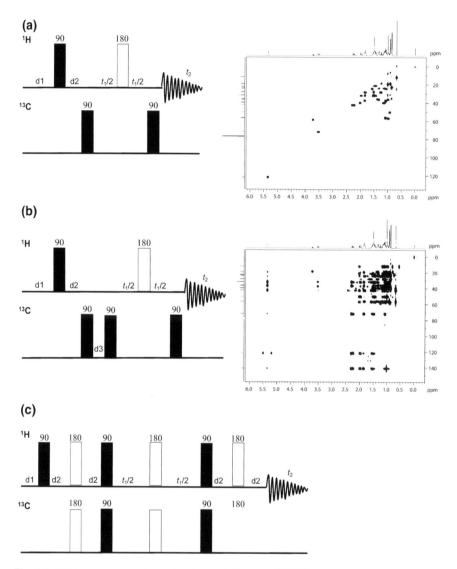

Fig. 3.6 2D heteronuclear-single bond correlations. **a** HMQC sequence and spectrum of cholesterol; d2 is optimized to $1/2^1J_{CH}$. **b** HMBC sequence and spectrum of cholesterol; d2 is optimized to $1/2^1J_{CH}$ and d3 to $1/2^nJ_{CH}$. **c** HSQC sequence: d2 is optimized to $1/4^1J_{CH}$

the experiment, as the diversity of couplings contributing to the FID makes it difficult to attain a full refocusing of the magnetization and the decoupling would lead to severe loss of information.

The HSQC (Heteronuclear Single Quantum Coherence) uses a slightly different approach to achieve single-bond correlations between ^{1}H and any heteronucleus (Fig. 3.6c) (Bodenhausen and Ruben 1980). The HSQC scaffold might initially appear more complex than the HMQC experiment, but it can be deconvoluted into INEPT and retro-INEPT blocks separated by a $180°$ pulse in the middle of the evolution time. The starting INEPT block generates single quantum coherences (SQC) between ^{1}H-X that evolve during t_1. An important characteristic of the SQC is that no homonuclear ^{1}H-^{1}H couplings can evolve, in contrast to what happens in the HMQC. The chemical shift of the heteronucleus X also evolves during t_1, while the ^{1}H-X coupling is refocused by the $180°$ pulse. The HSQC cross peaks are not split or broadened by homonuclear couplings and can be detected in phase-sensitive mode; therefore, in terms of resolution it is superior to the HMQC experiment. This is of particular interest in large molecules with very overcrowded spectra, such as biomolecules. In fact, the ^{1}H-^{15}N HSQC experiment is almost mandatory in any protein study by NMR.

In both the HMQC and HSQC experiments, the initial excitation (first radio-frequency pulse) of the magnetization is always performed on the more sensitive of the two nuclei, normally the ^{1}H. The final detection is on the same nucleus, achieving very high sensitivity gains, especially if very-low gamma X-nuclei are to be observed. However, ^{1}H is not the only possibility for a magnetization source and, if needed, the transfer of polarization can be performed via other nuclei, preferably of high sensitivity like ^{31}P or ^{19}F. Apart from the sensitivity gain, the design of both sequences situates the heteronucleus chemical shift range in the f_1 dimension, which necessitates fewer acquisition points. This set-up is more convenient than the inverse because the X signals are generally well dispersed, requiring less resolution to define the resonances properly. This leaves the f_2 dimension with a larger allocation of points for the ^{1}H signals, which are usually crowded. The kind of information heteronuclear correlation experiments provide and the differences they present in terms of signal resolution is shown in several of the references (Norwood et al. 1990; Bax et al. 1990).

3.5.3 Double Resonance Experiments: Homonuclear Spin Decoupling; Heteronuclear Double Resonance and Broadband Decoupling

In Chap. 2 we described the concept of decoupling (Sect. 2.2.4) and we have used this notion in previous pulse sequence descriptions. Suppression of the scalar J coupling between spins by the application of a very strong radiofrequency field achieves a single line resonance. Decoupling is commonly used in ^{13}C experiments and heteronuclear correlations to increase the sensitivity of the signals by collapsing multiplets, but the application of selective homonuclear ^{1}H-^{1}H decoupling can also be used to identify connecting protons. In this case, irradiation of an individual multiplet in the ^{1}H spectrum will eliminate the splitting of that signal as well as corresponding couplings of any other ^{1}H resonances to which it is connected (see, for instance, the spectrum of ethanol in Fig. 2.1). This experiment thus identifies proton–proton connections, although its application for molecular characterization has been superseded by techniques like COSY/TOCSY. Nowadays homonuclear decoupling is mainly applied to simplify multidimensional spectra and increase S/N. Experimentally, decoupling is achieved by irradiating at a specific ^{1}H frequency with a low-power, restricted frequency-range pulse known as a *soft* pulse, while the spectrum is acquired using the typical *hard* radiofrequency pulse. Two ^{1}H irradiation frequencies cohabit in the same experiment, giving rise to the name *double resonance*: one generating the full spectrum, applied in the middle of the chemical shift range, and a second selective frequency applied at the specific signal whose multiplicity, and that of its coupled partners, we aim to suppress.

Analogous to the selective decoupling we have just described, it is possible to achieve a decoupling scheme that affects a whole range of frequencies for a given nucleus. This is useful when no structural or coupling information is sought, but rather a raw increase in the simplicity and intensity of the resonances. Removing the splitting of signals simplifies multidimensional spectra dramatically, which is particularly important if there is a risk of cross-peak overlap. By applying the decoupling, a general increase in sensitivity is achieved via two mechanisms: the collapse of multiplets into singlets and the NOE effect between spins (Chap. 2). Experimentally, decoupling consists of a *train* of pulses applied in a very short time. Modern decoupling schemes have been designed to compensate for B_1 miscalibration and are especially robust to non-ideal conditions so they work well without particular set-up requirements. Depending on the nucleus to be decoupled, the extent of the frequency range and the spectrometer field, several multi-composite pulse sequences are usually applied. Among them, some of the most commonly used are WALTZ-16, GARP, MLEV-16 or WURST (Freeman 1997).

3.6 Correlations Through Space

So far we have introduced several NMR experiments based on scalar coupling connections transmitted via the chemical bond. However, coupling interactions between spins also exist that occur through space: the dipolar couplings. The nuclear Overhauser effect or NOE, introduced in Chap. 2, is a result of such dipolar couplings and is at the basis of many NMR experiments. The importance of the NOE in structure elucidation by NMR is unique, since it informs about the three-dimensional molecular geometry. In this section, we describe two different methods to observe the NOE: steady-state and kinetic NOE. Which one to choose will be dictated by the size of the molecule under study (Neuhaus and Williamson 1999).

3.6.1 Steady-State NOE

In the simplest NOE 1D difference experiment (Fig. 3.7a), firstly we irradiate a ^{1}H signal to reach saturation and observe the effects on all the protons dipolarly coupled to it (i.e. spatially close). The signal saturation is attained by irradiating selectively at the frequency of the chosen signal with a continuous wave of radiofrequency instead of a hard pulse, and measuring the effect thus produced with the acquisition of a ^{1}H spectrum by the application of a 90° pulse. In a second accompanying experiment, the pulse sequence is repeated but shifting the frequency of irradiation to a region of the spectrum without signals, so no NOE effect will be produced and the ^{1}H will show no signal enhancement. The difference between these two 1D spectra will yield only those resonances that see an enhancement in their intensities due to the initial NOE effect. In practical terms, this experiment is very useful to resolve the topology of a molecule and identify which protons are close in space. It can also provide structural information between fragments of a molecule that are not connected via the scalar-coupling based experiments.

This *steady-state* NOE (η) (from the steady state population distribution created via cross-relaxation of dipolarly coupled spins counteracted by their return to equilibrium) is dependent on the γ of the nuclei involved $\eta_{max} = 1/2(\gamma_A/\gamma_X)$, a characteristic that becomes relevant when the interacting spins are different. Although structurally-aimed heteronuclear NOE experiments are applied to some extent, the heteronuclear NOE is normally used as a sensitivity enhancement

Fig. 3.7 Correlations through space. **a** Pulse sequence of the NOE 1D-difference experiment. **b** Pulse sequence of the homonuclear 2D NOESY experiment

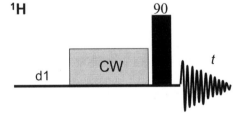

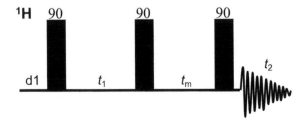

method via broad decoupling in experiments such as 1D ^{13}C, DEPT or those described in Sect. 3.5.3. Finally, it is important to bear in mind that the sign of γ is relevant in the above discussion, and therefore the NOE effect will be negative for nuclei of $\gamma < 0$ (e.g. ^{15}N, ^{29}Si, ^{119}Sn, etc.). Depending on the size and molecular tumbling of the molecule under study, a η_{max} near -100 % could be achieved for those $\gamma < 0$ nuclei, rendering a null signal instead of the NOE enhancement expected.

3.6.2 Kinetic or Transient NOE

NOE enhancement can also be achieved by initially creating a general perturbation on a molecule via radiofrequency pulses and leaving the system to evolve in the absence of any other perturbation. In this way, the spins will relax with each other and the NOE progressively builds up until a maximum is reached from where it decays back to zero. This *kinetic* NOE experiment, also called *transient* NOE, is the foundation of the NOESY experiment shown in Fig. 3.7b. The first two 90° pulses separated by the evolution time t_1 create the conditions for the spins to be aligned on the z-axis, and during mixing time, t_m, they will relax with each other and give rise to the NOE build-up. With the final 90° pulse the magnetization is returned to the xy-plane where it can be acquired. The resulting spectrum will show all possible NOE correlations within a molecule in a single experiment (see Fig. 4.3c for the NOESY spectrum of a protein).

Molecules are multi-spin systems, and therefore, the NOE effect on a single proton cannot be isolated from all its surrounding atoms. This circumstance leads to the issue of *indirect* effects on the NOE: the NOE enhancement produced on nucleus B by the proximity of A will be conveyed to nucleus C that is close to B but not to A. An NOE will be observed between A and C, even though they are not close enough in space to produce it. These indirect effects between nuclei that are further apart than the maximum distance to generate NOE connections are a consequence of long mixing times in the pulse sequence. In the case of small molecules these indirect NOEs have different sign to the direct NOEs and can therefore be easily identified. However, for large molecules such as proteins there is no sign difference between direct and indirect NOEs, which can become a serious problem when performing internuclear distance analysis.

This indirect effect is also known as *spin diffusion*. Although it is almost unavoidable, its presence can be discerned by acquiring experiments with increasing mixing times, thereby presenting different NOE build-up rates (the shorter the mixing time the less spin diffusion we will observe). The rate of growth of the NOE is directly related to the internuclear distance between spins by an r^{-6} factor. If the intensity of a specific NOE signal is linked to a known internuclear distance in the molecule (e.g. based on an X-ray structure or from an assigned secondary structure in a biomolecule), this cross peak calibration will give us a reasonable distance-range *versus* intensity correlation for the rest of resonances in the spectrum.

The relevance of this relationship between NOE intensity and internuclear distances will become clear in the next chapter, dedicated to biomolecules.

3.6.3 The 2D NOESY Sequence and Practical Aspects of the Experiment

The 2D NOESY experiment is normally applied to large molecules such as proteins or nucleic acids and follows the usual scheme for a multidimensional experiment. The mixing time during which the transient NOEs between spins develop is usually in the range of 50–300 ms, depending on the size of the molecule (as a rule of thumb, the larger the molecule the shorter the mixing time). The main features of the spectrum are similar to those seen for COSY or TOCSY, with the central difference that in the NOESY experiment the cross-peaks represent two nuclei located at distances no greater than 5–6 Å. Although spin diffusion is always present and needs to be taken into account, the intensities can be correlated with internuclear distances (r^{-6}), bearing in mind that the intensity of the NOE peak is contributed by all surrounding protons, with the nearest ones contributing to a larger extent.

As we already pointed out in Chap. 2, for medium-sized molecules η_{max} goes from positive to negative values in the region of intermediate molecular tumbling. The correlation time depends on the size and shape of the molecule, the viscosity of the solvent and the temperature of acquisition but there is a real risk that the NOESY experiment yields none or hardly any NOE enhancements for a significant range of molecules. In these cases the use of the alternative ROESY (Rotating frame Overhauser Effect SpectroscopY) sequence is recommended (Neuhaus and Williamson 1999). This experiment makes use of a spin-lock scheme like that applied in the TOCSY, but of weaker strength, that is able to generate a similar rotating frame Overhauser effect or ROE. Under these experimental conditions the NOE/ROE is always positive and the near-zero enhancement is no longer an issue.

References

Aue WP, Bartholdi E, Ernst RR (1976) Two-dimensional spectroscopy. Application to nuclear magnetic resonance. J Chem Phys 64:2229–2246

Bax A, Freeman R (1981) Investigation of complex networks of spin–spin coupling by two-dimensional NMR. J Magn Reson 44:542–561

Bax A, Freeman R, Kempsell SP (1980) Natural abundance C-13–C-13 coupling observed via double quantum coherence. J Am Chem Soc 102:4849–4851

Bax A, Freeman R, Frenkiel TA, Levitt MH (1981a) Assignment of carbon-13 NMR spectra via double-quantum coherence. J Magn Reson 43:478–483

Bax A, Freeman R, Frenkiel TA (1981b) An NMR technique for tracing out the carbon skeleton of an organic molecule. J Am Chem Soc 103:2102–2104

Bax A, Griffey RH, Hawkins BL (1983) Correlation of proton and nitrogen-15 chemical shifts by multiple quantum NMR. J Magn Reson 55:301–315

Bax A, Ikura M, Kay LE, Torchia DA, Tschudin R (1990) Comparison of different modes of two-dimensional reverse correlation NMR for the study of proteins. J Magn Reson 86:304–318

Bax A, Summers MF (1986) Proton and carbon-13 assignments from sensitivity-enhanced detection of heteronuclear multiple-bond connectivity by 2D multiple quantum NMR. J Am Chem Soc 108:2093–2094

Bax A, Subramanian S (1986) Sensitivity-enhanced two-dimensional heteronuclear shift correlation NMR spectroscopy. J Magn Reson 67:565–569

Buddrus J, Bauer H (1987) Determination of the carbon skeleton of organic compound by double-quantum coherence carbon-13 NMR spectroscopy—the INADEQUATE pulse sequence. Angew Chem Int Ed Engl 26:625–642

Berger S, Braun S (2003) 200 and more NMR experiments. Wiley-VCH, New York

Bodenhausen G, Ruben DJ (1980) Natural abundance nitrogen N-15 by enhanced heteronuclear spectroscopy. Chem Phys Lett 69:185–189

Braunschweiler L, Ernst RR (1983) Coherence transfer by isotropic mixing: application to proton correlation spectroscopy. J Magn Reson 53:521–528

Claridge TDW (1999) High-resolution NMR techniques in organic chemistry. Pergamon Press, Oxford

Cavanagh J, Fairbrother WJ, Palmer AG 3rd, Skelton N (1996) Protein NMR spectroscopy. Principles and practice, 1st edn. Academic Press, New York

Derome AE (1987) Modern NMR techniques for chemistry research. Pergamon Press, Oxford

Doddrell DM, Pegg DT, Bendall MR (1982) Distorsionless enhancement of NMR signals by polarization transfer. J Magn Reson 48:323–327

Ernst RR, Bodenhausen G, Wokau A (1987) Principles of NMR in one and two dimensions, Chapter 4. Clarendon Press, Oxford

Freeman R (1997) Spin choreography: basic steps in high resolution NMR. Spektrum, Oxford

Freeman R, Mareci TH, Morris GA (1981) Weak satellite signals in high resolution NMR spectra: separating the wheat from the chaff. J Magn Reson 42:341–345

Hoch JC, Stern AS (1996) NMR Data Processing. Wiley-Liss, New York

Jakobsen HJ, Linde SA, Sørensen S (1974) Sensitivity enhancement in ^{13}C FT NMR from selective population transfer (spt) in molecules with degenerate proton transitions. J Magn Reson 15:385–388

Jeener J (1971) Ampere International Summer school. Basko Polje, Yugoslavia

Lerner L, Bax A (1986) Sensitivity-enhanced two-dimensional heteronuclear relayed coherence transfer NMR spectroscopy. J Magn Reson 69:375–380

Morris GA, Freeman R (1979) Enhancement of nuclar magnetic resonance signals by polarization transfer. J Am Chem Soc 101:760–762

Muller L (1979) Sensitivity enhanced detection of weak nuclei using heteronuclear multiple quantum coherence. J Am Chem Soc 101:4481–4484

Neuhaus D, Williamson MP (1999) The nuclear Overhauser effect in structural and conformational analysis, 2nd edn. VCH Publishers, New York

Norwood TJ, Boyd J, Heritage JE, Soffe N, Campbell ID (1990) Comparison of techniques for proton detected 1H–15 N spectroscopy. J Magn Reson 87:488–501

Parella T (1999) NMR guide. Bruker, Karlsruhe

Rance M, Sørensen OW, Bodenhausen G, Wagner G, Ernst RR, Wüthrich K (1983) Improved spectral resolution in cosy 1H NMR spectra of proteins via double quantum filtering. Biochem Biophys Res Commun 117:479–485

Sørensen S, Hansen RS, Jakobsen HJ (1974) Assignments and relative signs of ^{13}C-X coupling constants in ^{13}C FT NMR from selective population transfer. J Magn Reson 14:243–245

Summers MF, Marzilli LG, Bax A (1986) Complete proton and carbon 13 assignments of coenzyme B12 through the use of new two-dimensional experiments. J Am Chem Soc 108:4285–4294

Chapter 4
Biomolecular NMR

Abstract One of the most relevant applications of NMR is in the study of biomolecules, which are at the heart of Biochemistry and Biomedicine. We shall describe in this chapter the use of NMR in the biomolecular field, especially its contribution to the structural elucidation of proteins and nucleic acids. An introduction to the analysis of biomolecular dynamics by NMR will also be described, as well as NMR applications in drug discovery and biomolecule-ligand interactions in general. Finally, the basic concepts behind solid-state NMR and metabolomics-by-NMR will be presented.

Keywords Affinity · Biomolecule · Drug-screening · Dynamics · Magic angle spinning · Metabolomics · NOE · Nucleic acid · Protein · Three-dimensional structure · Triple resonance experiment

4.1 Introduction

The application of NMR techniques to the study of biomolecules has marked a remarkable change within the whole field of Biochemistry in the last three decades. Being able to study the structure, molecular interactions and dynamics of proteins and nucleic acids in solution constitutes an excellent approximation to the actual working conditions in the cell and opens up the possibility of achieving the study of their behaviour in real time. Before the advent of biomolecular NMR only X-ray crystallography was able to produce structures at atomic resolution of biomolecules. However, the traditional bottleneck of X-ray is the crystallisation process, as not all molecules crystallise or give crystals of enough diffraction quality, while NMR does not suffer from that problem. Also, being able to study a biomolecule in solution is an advantage when trying to reproduce in the laboratory experimental conditions similar to the cell environment or in order to manipulate the biomolecule's behaviour by modifying its chemical surroundings. A good

R. J. Carbajo and J. L. Neira, *NMR for Chemists and Biologists*,
SpringerBriefs in Biochemistry and Molecular Biology,
DOI: 10.1007/978-94-007-6976-2_4, © The Author(s) 2013

measure of the importance of NMR in the study of biomolecules accounts from the number of structure coordinates deposited in the structural repository of reference, the Protein Data Bank (www.pdb.org). Of the nearly 90,000 biomolecular structures deposited in the database (March 2013) approximately 12 % have been obtained with NMR methods, which is quite a feat considering biomolecular NMR is still very young.

The major breakthrough that revolutionised the field took place in the early 1980s in Wüthrich's group at the ETH-Zurich. Their demonstration that the structure of a protein could be elucidated by NMR methods started the burgeoning field of biomolecular NMR. It brought a legion of biologists and biochemists to a research area that had been to that point the exclusive domain of physicists and chemists (Wüthrich 1986). At the same time, the expansion of NMR into biological areas sparked the development of new NMR techniques and methods specially adapted to the particularities of biomolecules such as their high molecular weights, methods of production, concentrations and possibilities of isotopic labelling. Thus, many advances in NMR in the last few decades have come from its application to biological problems. In this chapter we shall mainly present applications specifically designed for the study of such biomolecules in the fields of structural biology, drug discovery, metabolomics and solid-state NMR.

4.2 Why Biomolecules? Main Applications

The application of NMR techniques to the study of biomolecules can be conceived as a scaling-up from methodologies used for small molecules. The growing interest in the NMR of biomolecules is the main driving force behind the development of higher field spectrometers. All the typical biomolecules (proteins, nucleic acids or polysaccharides) are formed from repeating units of chemically-bound building blocks: amino acids, nucleotides or monosaccharides. Each of these building blocks is a simple organic molecule and its structural assignment can in principle be obtained in a straightforward manner with the 1D/2D NMR experiments described in the previous chapters: ^{1}H, ^{13}C, COSY, TOCSY, HSQC/HMQC, etc. At this point, the obvious question we could ask ourselves is why do we need a whole separate chapter to describe specific applications for biomolecules? The quick answer is that biomolecules are large or very large chemical entities and molecular size has a big influence on the magnetic behaviour as well as leading to a significant overcrowding of resonances in the spectra. To illustrate this size difference, a medium-sized molecule like cholesterol with a mass of 389 Da already presents a complex ^{1}H NMR spectrum (Fig. 3.2a). If we compare it with the ^{1}H spectrum of a 100 amino acid protein (considered small by biology standards, molecular weight around 11–12 kDa), it will present a 30× increase in the number of signals in the NMR spectrum (Fig. 4.1). Thus, the first problem we shall encounter when dealing with a biomolecule by NMR is the vast number of resonances in the ^{1}H, ^{13}C or ^{15}N spectra and even at very high-fields signal overlap

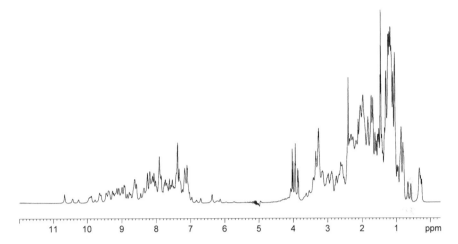

Fig. 4.1 1D-^{1}H-NMR spectrum of a protein. 1D ^{1}H spectrum of a 12 kDa protein (115 aa), showing the aliphatic (*right*) and NH/aromatic regions (*left*)

is a major issue. As we saw in Chap. 3 for the structural characterization of small molecules, the information provided by 1D experiments is commonly not enough to achieve a complete molecular assignment, normally requiring the acquisition of 2D homo- and heteronuclear spectra. In the case of biomolecules, 1D experiments are of hardly any use and 2D spectra have a limited applicability as they too present a significant cross-peak overcrowding.

The size of the biomolecule is the major culprit of this resonance congestion, but the repetition of chemical subunits (20 amino acids, 4 nucleotides and mainly two sugar-rings) that form proteins, nucleic acids and polysaccharides also share the blame. Consider for instance a protein that contains five alanine amino acids in its primary sequence. The chemical structure of this particular amino acid (NH, CO, CHα, CH$_3\beta$) will be the same for the five residues independent of their position in the sequence, resulting in very similar chemical shifts for any of the proton and carbon signals. However, chemical shift differences will normally be observed between those alanine residues arising from alterations in their respective chemical environments due to (i) the primary structure of the protein (i.e. what amino acids are chemically bound in the sequence through the peptide bond), (ii) the secondary structure of the protein fragment where that particular amino acid is located (the angles between the peptide bonds will vary accordingly to their location in an α-helix, a β-sheet, a loop or a non-structured or random coil region) and (iii) its tertiary structure, which determines what other amino acids in the sequence are spatially close. For example, the effect of the aromatic ring of a phenylalanine close to any of the alanine residues will be very different from that of a glycine which lacks a side-chain moiety. The main consequence of the combination of all these factors is that the expected signals for an alanine in the

^{1}H spectrum (NH, CHα, CH$_3\beta$) will normally show poor chemical shift dispersion among them. The above argument is valid for each of the other 19 amino acids that constitute proteins. It gets even more crowded for DNA and RNA where the chemical variability is much smaller with only four nucleotides contributing to the structures, or even worse for polysaccharides that are mainly constituted of two sugar-rings. This high complexity of biomolecule signalling in small chemical shift spaces drives the demand for application of multidimensional (3D or even added dimensions) experiments for their NMR study.

The acquisition of n-dimensional spectra is not the miracle cure for the NMR study of biomolecules; there are other sample requirements that need to be considered. For example, molecule concentration is an important issue to take into account. While the NMR samples of small molecules can be normally prepared to any desired concentration (unless they are of limited availability), in the case of biomolecules there are concentration limits specific for each molecule. Due to their nature and size it is commonly very difficult to achieve concentrations of over 1–2 mM, with a high risk of aggregation and consequent precipitation in the NMR tube. Although these concentrations are sufficient for ^{1}H detection, in the case of the ^{13}C nuclei only 1 % of the sample contributes to the NMR signal (Table 1.1), equivalent to 10 µM concentration in a 1 mM sample, making signal detection in a reasonable time a concern. This concentration problem is worse when measuring the ^{15}N nucleus, due to its intrinsically lower natural abundance and sensitivity.

In protein NMR, small polypeptides (up to 10 kDa) can be studied in their natural abundance form if the overlapping of resonances does not result in an insurmountable task. Over that 10 kDa size limit biomolecular NMR at natural abundance becomes an extremely complex or even impossible job. Chemical synthesis of proteins would make it possible to incorporate isotopes but has problems such as the size limit it can achieve (up to 70–100 aa), purity, yield and total cost. To surpass this big hurdle isotopic labelling of proteins was introduced. In a nutshell, making use of recombinant methods the DNA sequence coding for the protein of interest is introduced in the DNA of a host (normally *E. coli*, although other production methods are available (Mossakowska and Smith 1997; Graslund et al. 2008)), and the bacteria are "fed" with isotopically labelled nutrients such as ^{13}C-glucose and ^{15}N-ammonium chloride (for more details on the expression and purification of proteins, the reader is directed to specific texts on biochemical methods (Mossakowska and Smith 1997)). In this way all the products generated by the cell for its own subsistence as well as the target protein will be isotopically labelled. Once this protein is purified from the rest of debris/ molecules in the cell, it should present the same chemical characteristics and biological behaviour of a protein extracted from its natural source, but with nearly all the carbon and nitrogen atoms 100 % NMR-active. Under these conditions and with enough protein concentration, multidimensional experiments become feasible in principle. Even with good concentrations and labelling, stability of samples is another issue to bear in mind when working in biomolecular NMR, as the acquisition of all the needed multidimensional experiments can take from a few days to several weeks. For short-lived samples (and many biomolecules are!), instability becomes a big concern at the time of experimental acquisition. Finding ways to improve the stability of the sample via the production, purification or the addition of chemical stabilizers avoids the requirement of fresh protein samples with the consequent reduction in laboratory time and equipment costs.

As described previously (Chaps. 2 and 3) the correlation time, τ_c, of a molecule has a large influence on the NOE effect as well as in the relaxation properties of the spins. Small molecules show very fast tumbling and the effect on the relaxation is not apparent in the spectra, but this trend changes for medium to large biomolecules, where their sizes, and consequently slower rotations, bring important effects to spectra measurement. In fact, these relaxation properties are largely responsible for the size-limit in the study of proteins: as the size increases so does the τ_c, causing a reduction in the T_2 times of the signals (Chap. 1); shorter relaxation times mean fast signal decays, therefore low-information FIDs and broad signals. Short-lived magnetization in a context of long multidimensional pulse sequences with multiple delays and pulses will result in hardly any detectable signal at the end of the experiment. Although implementations have been introduced to overcome this restriction (such as the TROSY sequences (Riek et al. 2000; Kay 2011a; Tugarinov et al. 2004) or selective labelling of biomolecules (Ruschak et al. 2010; Religa and Kay 2010; Lundström et al. 2009; Gardner and Kay 1998)), the size of the protein or nucleic acid is still the main limitation in the field. Whereas biomolecules up to 25 kDa can be studied using standard experiments, above this molecular weight it becomes increasingly difficult to get high resolution structures. However, the target in biomolecular NMR is not necessarily always a high resolution structure of the molecule, as other useful biochemical information can be extracted via NMR methods regardless of the size: protein/nucleic acid interactions, applications in drug discovery, dynamics, etc. Some of these approaches are described in more detail in the following sections of this chapter.

In the field of biomolecular NMR the use of high magnetic fields is especially relevant. Firstly, due to the increase in signal sensitivity that can be obtained for low concentration samples (micromolar range), but most importantly for the resolution enhancement high fields produce that allows the differentiation of signals in overcrowded NMR spectra. As a rule of thumb, a minimum spectrometer field of 500 MHz will be necessary for the study of biomolecules, and even higher fields are recommended for the acquisition of interpretable homonuclear ^1H experiments (TOCSY, NOESY). Proteins, nucleic acids and sugars in general present a characteristic high solubility in water. Therefore, biomolecular NMR is generally performed in aqueous buffers to control the pH, with the addition of a 5–10 % of D_2O for lock purposes. A 100 % D_2O solution is not normally used because biomolecules contain numerous exchangeable protons that would be lost, such as NH groups in amino acids and nucleotides. These NH groups provide basic information for the structural study of the biomolecule and need to be retained. However, the use of water as solvent brings about a ^1H signal for H_2O that dominates the NMR spectra due to its high concentration in the sample (pure water is 55 M compared to an ideal mM range sample). Several water suppression schemes have been developed for the elimination or substantial reduction of the water signal, such as presaturation at its chemical shift frequency or the WATERGATE and jump-and-return pulse blocks (Price 1999). These schemes are so important that a good implementation of the water suppression tools can be decisive in the quality and final results achieved from biomolecular NMR experiments.

4.3 Structure of Biomolecules

The primary structure of the biomolecules (i.e. their amino acid sequence) is normally known in advance at the start of any structural study: if it is obtained via recombinant methods the DNA and therefore the amino acid sequence is established a *priori*. However, in the case of proteins from natural sources, NMR studies are known to have corrected the previously attributed sequence. This information, together with knowledge of the individual chemical structures of the 20 natural amino acids or the four nucleotides, is a good starting point for the analysis of any protein or nucleic acid. The final aim of a full structural study is the determination of the secondary and tertiary structure of the molecule by establishing the spatial localization of each of those amino acids/nucleotides with respect to each other.

To calculate a biomolecular NMR structure you input the known chemical structure (from the amino acid sequence in the case of a protein) in a random orientation into a molecular dynamics calculation protocol. This calculation constrains the structure using all the information you have gained regarding chemical bond connections determined by the scalar couplings (homonuclear and heteronuclear) and spatial relationships obtained from dipolar couplings via NOE-based experiments. There are also orientational restraints provided by residual dipolar couplings (RDCs) that give information about the angle relative to the external magnetic field (Lipsitz and Tjandra 2004). These RDCs inform about the relative orientation of parts of the molecule that are far apart in the structure, which is especially relevant in large molecules (>25 kDa) as it is often difficult to record NOEs due to spin diffusion. You then repeat the calculation with many different random orientations, generating an ensemble of structures that comply with those geometrical restrictions and the inherent chemical characteristics (bonds, torsion angles, van der Waals radii) of the molecules. The calculation of a molecular structure by NMR is an iterative refinement process: with each structural outcome new restraints are incorporated to the calculation that conform to the most recent structures as well as reassignments of restraints that do not agree with them (structural violations). At the end of the assignment and calculation procedures the structures will converge to be very similar to each other as measured by their backbone root mean square deviation or rmsd, where a value below 1.0 Å is normally considered a well-defined NMR structure. Generally, more information will lead to better defined structures.

The first step is to know which nuclei each of the signals arise from. This usually long and complex procedure is commonly known as the *assignment* process, and it will be briefly described next for the case of a protein. Assigning a large molecule like a protein by NMR is like solving a jigsaw puzzle formed by several thousand pieces. However, instead of reassembling a picture that has been cut into many interlocking pieces, we shall manage thousands of cross peaks (the pieces) generated from complementary NMR spectra, that allow the identification of the resonances for each of the amino acids in the sequence and their interactions in space. Once all the pieces have been put together, the result will be the complete three-dimensional structure of the protein (the final picture). Although in the case of a previously unsolved protein there might be no previous knowledge of what the final picture will look like, some partial information can be gathered from homologous proteins, structural prediction algorithms, etc. In addition, some pieces of our three-dimensional model may be missing if their corresponding molecular fragments do not show a good behaviour for the NMR (for instance, due to fast relaxation, broad signals or severe overlapping), but if only small regions are affected, the global picture of the molecule should be almost complete.

To solve a jigsaw puzzle one usually starts with those pieces easy to locate because they show some particularities that distinguish them from the rest. Likewise, a similar strategy is followed in the NMR assignment by trying to assign those resonances that, for different reasons, stand out, and from there build up the rest of the assignment, leaving those harder-to-assign resonances/pieces for the very end when not much is left to be assembled. For our biomolecular NMR jigsaw the pieces we need are the chemical shift values for each ^1H, ^{13}C and ^{15}N atom in the molecule so that we know which atoms each signal arises from. Making use of the combination of several multidimensional experiments, all the gathered information is used with the aim of obtaining structural information regarding distances between protons (mainly, but not exclusively through NOEs, Sect. 4.3.1), dihedral-angles between nuclei (from homo or heteronuclear correlation spectroscopy, Chap. 3), and chemical shift values.

4.3.1   Homonuclear and Heteronuclear (Triple Resonance) Assignment in Proteins

Tens if not hundreds of experimental set-ups have been developed for use in biomolecular NMR. Deciding which one is best for a specific task can be far from easy for the non-specialist. In this section we shall focus on a handful of well-established methods which are commonly used to provide the basic information to conduct the assignment of a biomolecule. Among these experiments are the already described homonuclear sequences ^1H-^1H TOCSY and NOESY (Chap. 3) in their corresponding 2D and isotope-edited 3D versions. Other experiments are based on heteronuclear correlations involving the three main nuclei in biomolecules, $^1H/^{13}C/^{15}N$, within the same pulse sequence known as *triple-resonance* experiments (Kay et al. 1990). The construction of a 3D experiment follows the scheme already described for a 2D spectrum, with the incorporation of a second evolution period with a gradually incremented delay during the acquisition process. This second evolution time will generate the third FID necessary to render the final 3D spectrum. As for 2D experiments, total acquisition time is determined, apart from the number of scans or repetitions, by the number of points allocated for each dimension. The addition of a third dimension and the accompanying points to define it, will dramatically increase the total acquisition time of the experiment. To collect a 3D spectrum for a protein using eight scans for each FID, 128 points for the second dimension and 100 for the third, we will end up collecting a total of $8 \times 128 \times 100 = 102{,}400$ FIDs. If the average time to collect one FID is 3 s, a 3D experiment will take three and a half days of acquisition time. 3D spectroscopy is far from trivial in terms of measurement time, especially when several of those experiments have to be acquired. This also has implications for sample stability. A selection of the most important protein NMR experiments will follow, with a brief description of the information they provide.

4.3.1.1 ^{15}N-HSQC

The HSQC pulse sequence has been described in some detail in Chap. 3. The ^{1}H-^{13}C version is routinely applied to the characterization of small molecules and also used in proteins during the assignment process mainly to help in the determination of ^{13}C chemical shifts for each amino acid. However, the ^{15}N version is the one that has found wide applicability in the biomolecular field as it provides both qualitative and quantitative information about the protein. This heteronuclear ^{1}H-^{15}N correlation experiment connects N–H nuclei through their ^1J scalar coupling (Fig. 4.2). In the chemical structure of proteins, there is an N–H bond per amino acid in the peptidic backbone (except in the case of the amino acid proline which lacks that bond due to its cyclic nature). Therefore we expect the HSQC of any protein to contain as many cross peaks as amino acids in the sequence minus the number of prolines. Additional signals arise from those amino acid side chains that contain NH/NH$_2$ groups: glutamine, asparagine, arginine and lysine. In the case of the NH$_2$ groups, a unique ^{15}N chemical shift in the nitrogen dimension will correlate with two ^{1}H differentiated resonances. Finally, the NH group from the indole ring of tryptophan gives rise to an additional cross peak in the HSQC spectrum.

The ^{15}N-HSQC spectrum is known as the *fingerprint* of a protein, as this map of cross peaks ideally shows a well dispersed pattern of signals that is easily recognisable. This characteristic makes the HSQC experiment ideal for its use as quality control of protein production. Besides, it allows its application in the determination of protein–protein or protein–ligand interactions, as it will be described later (Sect. 4.5). The HSQC also provides qualitative information about the *folding* of a protein: a protein presenting an HSQC with fewer cross peaks than expected or deficiently resolved indicates a partially unstructured molecule or the formation of aggregates in solution (Fig. 4.2b). In contrast, a well-dispersed

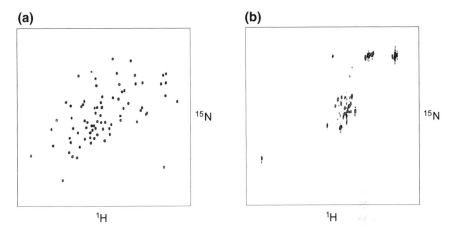

Fig. 4.2 ^{15}N-HSQC experiment in proteins. **a** The ^{15}N-HSQC spectrum of a 72 aa, well-folded protein. **b** The ^{15}N-HSQC spectrum of a 70 aa unfolded protein under native conditions

^{15}N HSQC accounting for all the expected signals will be proof of an adequately folded protein (Fig. 4.2a).

NH protons chemically exchange with water (are labile), i.e. if they are not chemically or sterically impeded N–H bonds can be broken and formed again with protons from H_2O. To use this chemical property of the NH group to our advantage, experiments can be carried out in the presence of 100 % D_2O instead of H_2O. Those NHs accessible to the solvent because they are on the protein surface will disappear promptly from the HSQC spectrum, while those forming strong hydrogen bonds with other chemical groups in the protein, or not accessible because they are buried in the hydrophobic core, will experience a delay in their exchange with deuterium or not taking place (Sect. 4.4.2). This hydrogen-bonding information can be transformed into distance/bonding restraints in the structure calculation process, similar to a covalent bond. However, the full assignment of the ^{15}N-HSQC (i.e. correlating each cross peak with an amino acid in the sequence) can only be performed in combination with other complementary experiments, some of which will be described below.

4.3.1.2 2D/3D TOCSY

This spin-lock based experiment that was introduced in Chap. 3 depends on the relay of scalar couplings between protons within each spin system. In the case of proteins, the partial double-bond character of the peptide CO–NH bond disrupts inter-residual ^{1}H-^{1}H scalar coupling connections between protons of consecutive amino acids; therefore each residue presents an isolated ^{1}H spin system. Thus, the TOCSY experiment applied to proteins will show only connections within each of the amino acids in the sequence. For example, continuing with the amino acid

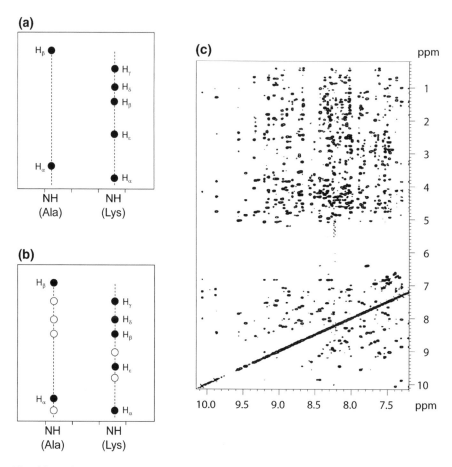

Fig. 4.3 Assignment via TOCSY/NOESY experiments. **a** 2D TOCSY cross peaks expected for the NH protons from alanine and lysine residues. Only intraresidual signals within each spin-system can be observed in the TOCSY experiment. **b** 2D NOESY: intra and interresidual cross peaks are observed. Their identity is established by comparing them with the cross peaks observed in the TOCSY for the same spin system. Apart from the intraresidual cross peaks, the NH of an amino acid is expected to yield NOE correlations with the side-chain protons of the preceding residue in the sequence. **c** NH/aromatic region of a 2D NOESY from a 120 aa protein

alanine, if we start from its NH proton it should show two cross peaks with the Hα and CH$_3$β protons, respectively. Equivalent cross peaks are found if we look at the connection from the Hα chemical shift or the CH$_3$β (Fig. 4.3a). If we consider a longer side chain like that of lysine, the spin-lock should be able to connect the whole spin system, from the N–H to the protons at α, β, γ, δ and ε positions in the side chain. In the aromatic amino acids (histidine, phenylalanine, tyrosine and tryptophan) the ^{1}H spin system cannot be completely relayed via scalar couplings within the side chain. The NH, Hα and Hβ protons are isolated from the aromatic ring protons due to a quaternary carbon, and no connectivities between both moieties are observed in the TOCSY spectra.

The assignment of each spin system in the TOCSY to the corresponding amino acid is based on the number and pattern of peaks observed and their chemical shifts. Each residue presents a typical range of frequencies depending on the nature of their side-chains. While some residues like glycine, alanine, threonine, isoleucine or leucine present distinctive spin-systems and chemical shifts, others like aspartate, asparagine, glutamine, glutamate or methionine show side-chain ^{1}H signals in a common range of frequencies. This ambiguity in the spin-system identification cannot be unambiguously solved solely with the TOCSY and complementary experiments such as NOESY are required. Also, the TOCSY experiment is able to assign the type of amino acid but it cannot determine which one of that type it is within the protein sequence due to the lack of inter-residual connections.

Significant improvements over the 2D TOCSY methodology are the isotopically-edited ^{13}C and ^{15}N 3D HSQC-TOCSY versions of the pulse sequence. An easy way to visualize this 3D experiment is to consider it a combination of the 2D ^{15}N-HSQC and ^{1}H-^{1}H TOCSY sequences, where the two-dimensional map of cross peaks of the TOCSY is dispersed within a cube where the third dimension is the ^{15}N chemical shift of the NH group. This 3D experiment allows the correlation of the ^{1}H spin system with its own nitrogen-15 atom, which can then be assigned to a particular type of amino acid in the ^{15}N-HSQC, information that in conjunction with other experiments will lead to the sequential assignment of the protein. The ^{13}C edited 3D TOCSY-HSQC pulse sequence follows the same arrangement but in this case the spin systems are discriminated in the third dimension according to the ^{13}C chemical shifts of the side chains.

4.3.1.3 2D/3D NOESY

This dipolar-coupling based experiment provides the most significant data for the determination of a biomolecular structure by NMR: spatial interaction. As already described (Chaps. 2 and 3), there is a direct correlation between the intensity of a NOE cross-peak and the interatomic distance of the spins that originate it, up to a 5–6 Å limit. The determination of any 3D structure is based on the establishment of relative spatial coordinates for each of the atoms that constitute the molecule, therefore, if we are able to determine what ^{1}Hs are within 5 Å of distance of a particular proton and repeat the same process for all hydrogen atoms we shall, in principle, have enough data to build the 3D structure of the biomolecule. Each of the ^{1}Hs in a protein will typically be in the vicinity of a handful of other ^{1}Hs located at the minimum distance to see the NOE effect which multiplied by the number of protons in the structure translates into hundreds or even thousands of cross peaks present in the 2D NOESY, becoming a very complicated jigsaw in its own right. The ^{1}H-^{1}H 2D NOESY complements the TOCSY experiment well for the spin system assignment by combining the intraresidual cross peaks within each spin system afforded by the TOCSY with the intra- and interresidual cross peaks from the NOESY. Cross peaks in the NOESY arise regardless of their origin from dipolar connections within the spin system or from a sequentially or spatially close amino acid, but the comparison with the TOCSY data will allow for that critical discrimination (Fig. 4.3b).

The peptide backbone connectivity characteristic of proteins results in the side-chains of sequential amino acids being spatially close to each other. Of special interest for our assignment purpose is the proximity of the amide NH proton of each amino acid to the side-chain of the preceding residue. This kind of sequential information is crucial for the assignment process of a protein, as it connects consecutive amino acids. A depiction of how the sequential assignment of a protein can be performed using both the NOESY and TOCSY experiments is shown in Fig. 4.3. Starting from the NH proton of an alanine residue (spin system assigned via the TOCSY, Fig. 4.3a), we observe several correlations with other ^{1}Hs. Those arising from its own side-chain (intraresidual NOEs) are identified by comparison with the cross peaks from the same NH in the TOCSY spectrum. The extra cross peaks of the alanine NH detected in the NOESY share the same chemical shifts of a lysine spin-system also previously identified, but they could have originated either because the alanine and lysine are sequential amino acids in the primary structure of the protein (sequential NOEs), or because they are sep-arated in the sequence but spatially close (long range NOEs). If the NH of that lysine correlates with the spin system of a residue X, and a fragment within the protein sequence corresponds to the tripeptide X-Lys-Ala, it is a good indication of a successful sequential polypeptide patch assignment. By adding new spin systems in both directions we expand the assignment of the protein, in the same way as the pieces of our jigsaw are joined together. Although assignment ambiguities or overlapping signals can at some point stop the sequential determination, the process can be resumed by finding a new starting point in the sequence. As with the TOCSY experiment, signal overlapping can be overcome with the use of 3D ^{13}C/^{15}N 3D HSQC-NOESY spectra which require isotopically labelled proteins.

4.3.1.4 HNCO

This 3D triple resonance experiment (Kay et al. 1990) combines resonances from the three nuclei involved in protein assignment, ^{1}H, ^{13}C and ^{15}N in the same pulse sequence (Fig. 4.4a) (Muhandiram and Kay 1994). In the ^{13}C frequency differen-tiated pulses are applied at the chemical shift ranges of both the ^{13}CO carbonyl and the ^{13}Cα regions. Common to many nD experiments, this experiment is built from INEPT/HSQC building blocks. In a nutshell, the HNCO transfers magnetization from the H nucleus of the amide group of each amino acid to its directly bound amide N atom via the $^1J_{NH}$ scalar coupling. This nitrogen atom then correlates with the ^{13}CO carbonyl of the preceding residue via the $^1J_{NCO}$ scalar coupling (which is very large ≈ 140 Hz). Lastly, for maximum sensitivity, magnetization is returned to the ^{1}H channel where it is detected in the end. The outcome of the HNCO experiment is the sequential correlation of each of the NH protons with its own nitrogen and the [13]CO carbon of the preceding residue (H → N → CO).

4.3.1.5   HNCA

This experiment is analogous to the HNCO (Kay et al. 1990; Grzesiek and Bax 1992), but provides connections between the amide NH group of the protein backbone and the α-carbons of both the preceding amino acid (via $^2J_{NC\alpha}$; weak signal) and its own residue (via $^1J_{NC\alpha}$; strong signal), combining spin-system and sequential information in the same experiment (H → N → Cα; Fig. 4.4b).

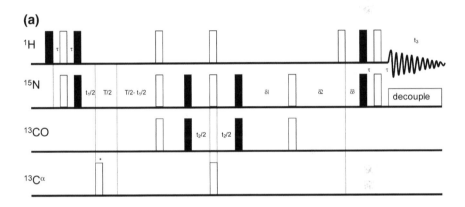

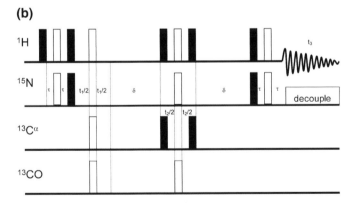

Fig. 4.4 Biomolecular 3D NMR. **a** HNCO in its *constant time* version; t_1 and t_2 are the incremented delays to create the second and third dimensions, respectively. **b** HNCA: the delays (δ and τ) are specified according to coupling constants $^1J/^2J(C\alpha$-N) and $^1J(N$–H) (see (Cavanagh et al. 1996) for further details); t_1 and t_2 are the incremented delays for the second and third dimensions, respectively. The phases of the different 90 and 180° pulses are also variable (Cavanagh et al. 1996)

Numerous triple resonance experiments are available (Fig. 4.5), and for ease of comprehension they are named according to the type of information they provide. For instance, the HN(CO)CA yields the correlation between the amide proton and the preceding α-carbon through the carbonyl group (and therefore between brackets). This experiment, in combination with the HNCA, allows distinguishing between intra-residue and preceding α-carbons. A more exhaustive list of triple-resonance experiments can be found in references (Riek et al. 2000; Kay et al. 1990). It is important to note that none of these experiments on its own allows assignment of the backbone or the side-chains of a whole protein, but it is the combination of the information each of them provides which affords the full assignment. Once the ^{1}H, ^{13}C and ^{15}N assignments have been obtained the NOE-based experiments can be fully interpreted, supplying interatomic distances (see above), while the chemical shift values and scalar couplings can be correlated with dihedral angle information. This spatial information is used in the structure calculation protocols to yield the ensemble of possible structures complying with the NMR data.

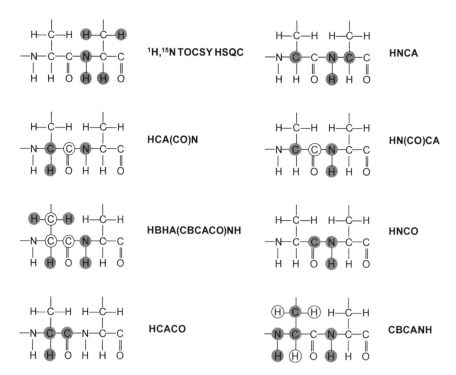

Fig. 4.5 Triple resonance experiments. The nuclei observed in the experiments are coloured; those nuclei which are not observed but allow transfer of magnetization are inside white circles. A more complete list of used 3D experiments in protein assignment is described in the literature (Cavanagh et al. 1996)

4.3.2 Nucleic Acids

The study of nucleic acids by NMR has seen significant improvements during the last twenty years in parallel with the evolution of the protein-NMR field. NMR spectroscopy is ideally suited to provide insights into the structure, dynamics, and function of nucleic acids, especially of small functional subunits. The introduction of isotopic labelling methods, first for RNA (Kime 1984) and more recently for DNA (Olsen et al. 1982; Patel et al. 1987), has paved the way for the application of triple resonance methods specifically designed for the nucleotides that constitute both classes of nucleic acids. The aim of this section is to outline some of the main aspects that need be taken into account when analysing polynucleotides by NMR. More comprehensive and thorough reviews on the state-of-the-art methodology for the assignment and structure of nucleic acids by NMR are available and the interested reader is directed to them (Neuhaus and Williamson 1999; Zerbe 2003).

The chemical structure of DNA constitutes a backbone of deoxy-ribose sugar rings bound by phosphate groups, with four different bases attached to the sugar: guanine (G), cytosine (C), adenine (A) and thymine (T). In the case of RNA the uracil (U) nucleotide replaces the thymine and the sugar is a ribose. Bases A and G are called purines, while C, T and U are pyrimidines. Hence, the diversity of chemical structures within nucleic acids is much smaller than that found in the 20 amino acids of proteins. This reduced structural variability translates into a reduction of chemical shift dispersion and consequently an increase of signal overlap in NMR spectra. This issue is especially important in the region where the sugar protons resonate, as only deoxy-ribose in DNA and ribose in RNA are present. Moreover, the chemical shift region where the sugar ^1H signals resonate is shared with the H_2O signal from the solvent, meaning good water suppression schemes are needed when dealing with nucleic acids.

Among the methods available to remove the water signal the most commonly applied in nucleic acids are jump-and-return and WATERGATE (Price 1999). Presaturation methods are avoided as they produce a significant loss of signal intensity in the labile protons due to saturation transfer to the fast exchanging imino/amino protons. 100 % D_2O samples are also of not much use as exchange with D_2O would eliminate the signals from the labile imino/amino protons of the bases, which is crucial for the assignment and structure determination of poly-nucleotides. As in protein NMR, the 1D ^1H spectrum works as a quality control of the DNA/RNA samples (folding, purity, homogeneity). The imino resonances which are generally located at low fields (above 10 ppm) provide qualitative information regarding the number of base pairs or the number of conformations present in solution.

For non-labelled samples of nucleic acids, the basic experiments used in their assignment are the homonuclear COSY and TOCSY, which provide independent connections for the ribose-base spin system characterization. The NOESY experiment renders cross-peaks which originate from intra-nucleotide interactions (common to COSY and TOCSY cross-peaks), from adjoining base pairs, sequential

NOEs and inter-strand ones. The latter are particularly relevant as they allow determination of the Watson–Crick base pairs, which are the basis of the double-strand structure of nucleic acids. These homonuclear 2D methods are often adequate for the study of small DNA/RNA units of about 20 nucleotides, or as initial experiments for larger systems. The assignment of nucleic acids follows similar procedures to those described for proteins, such as the establishment of good starting points from where to build-up the sequential connections that allow the linking of the nucleotides. Additionally, the presence of the phosphorus atom in the chemical composition of nucleic acids allows the incorporation of NMR experiments based on ^{31}P (spin ½, 100 % natural abundance, Table 1.2), making use of the dispersion of ^{31}P chemical shifts and correlation experiments based on ^{1}H-^{31}P and ^{13}C-^{31}P scalar couplings. There are ^{1}H-^{31}P correlations of the type HETCOR/HMQC and 3D ^{1}H-^{13}C-^{31}P experiments applicable for carbon-13 labelled DNA/RNA. Also for isotopically labelled nucleotides, other experiments are available with similar pulse schemes to those used for proteins: $^{15}N/^{13}C$-NOESY-HMQC or HCCH-COSY/ TOCSY (mainly for the ribose ring) (Cavanagh et al. 1996; Ikura et al. 1991; Olejniczak et al. 1992), the latter providing ^{1}H-^{13}C-^{13}C-^{1}H correlations via $^{1,2}J$ scalar couplings (COSY) or of the whole spin system (TOCSY). As in proteins, isotopic labelling extends the size of the systems that can be studied by the application of multidimensional experiments that solve the signal overlap. In addition, isotopically labelled samples allow the direct transfer of coupling information via the heteronuclei relieving the reliance on the ^{1}H-^{1}H connections of the homonuclear experiments, which are often ambiguous. Especially relevant are pulse sequences that correlate the base and ribose moieties via H–C–N scalar couplings, to solve the ambiguity between the intra/internucleotide cross-peaks from the NOESY experiments. For structure calculations for nucleic acids, distance restraints from the NOE-based experiments together with angle restraints via the measurement of J-couplings are used. Both homonuclear and heteronuclear scalar couplings are employed for the elucidation of ribose conformations (the sugar pucker) and glycosidic bond angles, as well as for establishment of backbone orientation via dihedral angles. For nucleic acid NMR experiments have been designed to detect J-couplings via the inter-strand hydrogen bonds formed between the base pairs (mainly H–N···H) which can later be introduced as semi-covalent restraints in the structure calculation protocols (Allen et al. 2001).

4.4 Biomolecular Dynamics

Proteins directly control or are involved in most biological processes. The common representation of the biomolecular structures obtained by NMR and X-ray depicts proteins as well-defined bundles of α-helices and β-sheets, but these do not render any sense of molecular motion. However, proteins and other biomolecules are intrinsically dynamic systems displaying different degrees of flexibility according to their structure and function. There is a strong relationship between the structure of a biomolecule, its dynamics and the function it performs in the cell. Examples of biomolecular dynamic processes are protein folding, protein–protein/DNA/RNA interactions, allosteric regulation or enzyme catalysis. These dynamic processes can take place in a restricted part of the biomolecule (local) or involve the whole system (global). In general any relevant interaction involving a biomolecule will result in some changes in its dynamics. For example, there is ample evidence indicating that target-binding protein sites are flexible spots. NMR is a very powerful biophysical technique to study the flexibility and dynamics of biomolecules as it is able to report on a wide range of time-scales. The theory of relaxation behind these studies (Chap. 1) is beyond the scope of this book (the interested reader can have a look at the literature (Cavanagh et al. 1996)). Here we shall focus on the main considerations to be taken into account, the possibilities that NMR offers to the study of different time-scales and the kind of information that NMR can provide regarding the intrinsic flexibility of biomolecules.

4.4.1 Comparison with Other Spectroscopies and Structural Biophysical Techniques

Molecular translation, due to the Brownian motion, can be measured by means of the translational diffusion coefficient D, which depends on the size and shape of the molecules that can be obtained from ultracentrifugation measurements (Tinoco et al. 2002). From these D values other parameters can be extracted such as the correlation time τ_c and the hydrodynamic radii of molecules. In this context, the rotation movement of proteins has been measured by polarization of fluorescence. In brief, a molecule can absorb light along certain specific directions depending on its molecular structure (Lakowicz 2010), and the corresponding fluorescence emission will be polarized. However, this emission decreases with the extent of molecular rotation experienced during the emission (in the order of 10^{-8}–10^{-9} s), thus allowing the measurement of the rotation of the molecule. Measurements of correlation times obtained by NMR or fluorescence have shown in general a good agreement, although some discrepancies have been found due to the influence of the water shell surrounding the molecules.

In some particular examples, deconvolution of the broad absorbance peak of proteins at 280 nm can yield some information about the dynamics of aromatic residues during the drastic events of protein folding and unfolding (Esfandiary et al. 2009) and the study of the infra-red region can yield some information on the dynamics of solvated biomolecules (Xu et al. 2006 and Arrondo and Goñi 1999). X-ray crystallography can also provide dynamical information through the B-factors which measure how good is the electronic density around a particular atom: the larger the B-factor, the higher the disorder (Petsko and Ringe 1984, 1985). However, it is important to indicate that all these techniques, except X-ray, provide overall information of the molecule as a whole, or of some specific residues (fluorescent residues or aromatic amino acids).

4.4.2 Movements in the ps-s Range

NMR spectroscopy allows the study of biomolecular dynamics in a range of time-scales spanning from seconds to picoseconds both globally and site-specifically (Fig. 4.6). The wide amplitude of timescales available opens the range of possible dynamic behaviours that can be studied, as different biological phenomena take place with different timescales. For example, dynamics corresponding with the motional amplitudes of amino acid side-chains or the flipping of aromatic rings occur at much faster rates than the overall tumbling of a biomolecule. Generally speaking, we can consider the dynamics of a protein as a combination of specific local dynamics taking place at the same time as the overall molecular tumbling of the biomolecule in the solution. Molecular tumbling is the main contributor to the reorientation of the N–H bonds which, although not exclusively, is the principal source of dynamic information within the protein for NMR spectroscopy (Chap. 1). The dynamics of the methyl groups have also been recently explored, providing clues on their mobility in the native state of proteins and on the presence of non-native interactions during the intermediates along the folding reaction (Religa and Kay 2010; Lundström et al. 2009).

4.4.2.1 Fast Protein Dynamics: ps to ns

To grasp the processes that nuclear spins undergo due to the flexibility of a molecule, we shall focus on a single N–H bond from the backbone of a protein, depicted as a vector forming an angle with respect the external magnetic field (Fig. 1.3). This bond will show different orientations for each of the molecules in the sample as they tumble freely in solution, and the intrinsic dipolar interaction existing between each ^{1}H and ^{15}N pair is averaged to zero in the whole sample due to this molecular tumbling. However, the local magnetic field that the N–H bond *experiences* will be perturbed by two main factors causing its reorientation: (i) the internal motions in the region spatially close to the NH bond and (ii) the general molecular tumbling of the protein. The fluctuations of the local magnetic fields are sensitive to internal motions. Therefore applying pulse sequences specifically designed to excite the H–N pairs in the molecule allows extraction of the NMR relaxation rates and analysis of dynamic parameters.

The NMR experiments employed for the measurement of the dynamics of backbone NH bonds are designed upon the building block of the 2D ^{1}H, ^{15}N HSQC experiment (Chap. 3). These experiments measure the transverse ($R_2 = 1/T_2$) and longitudinal ($R_1 = 1/T_1$) relaxation rates (Chap. 1) as well as the ^{1}H-^{15}N cross-relaxation rate measured via the ^{15}N{^{1}H} steady-state NOE (Chap. 2). Similarly to the typical HSQC heteronuclear correlation, a cross-peak is observed for each NH pair in the protein, where in this case the intensity is related to $\exp(-t/T_i)$ where T_i is the T_1 or T_2 value of the particular ^{15}N nucleus and t is a variable relaxation delay within the pulse sequence. To obtain relaxation rates, the HSQC is repeated several times (normally 10–12 spectra provide enough information) varying the relaxation delay t in each experiment. Once this series of spectra have been recorded and processed, peak intensities are fitted to an exponential as a function of t from which values of T_1 and T_2 can be extracted for each amino acid in the sequence.

The interpretation of these site-specific internal dynamics parameters in the sub-nanosecond timescale is normally done in terms of *order parameters* correlating with the amplitude of the motions of the NH bonds, and also with time constants that reflect the timescale of the internal motions. This information is obtained primarily using the *model-free* Lipari-Szabo formalism (so called because the parameters are derived without the need of a specific model for internal motion) (Lipari and Szabo 1982; Lipari and Szabo 1982). The model-free approach assumes that overall and internal motions are uncorrelated, which is normally true as the internal motions are generally small and do not affect the global diffusion of the molecule. In the model-free analysis the amplitude of the fluctuation of each NH bond (backbone dynamics), and therefore its site-specific flexibility, is given by the *order parameter* S^2 that ranges from 0 (unrestricted motion) to 1 (totally rigid). These nuclear spin-relaxation measurements are able to inform about fast (pico- to nanoseconds) and slow (micro- to milliseconds) timescales, as well as reporting about the overall rotational diffusion of the biomolecule which is in the ns timescale.

4.4.2.2 Slow Protein Dynamics: μs to ms

Conformational changes in this timescale alter the chemical environment of the spins, possibly modulating their chemical shifts (Fig. 4.6). If the differences in the chemical shift are large, the signals are broadened due to a decrease in $R_2 = 1/T_2$. R_2 rates are usually measured via spin-lock and Carr-Purcell-Meiboom-Gill (CPMG) relaxation methods (based on the spin-echo described in Chap. 1 to measure T_2 relaxation times). Motions in the ms-μs timescale are reflected in the NMR lineshapes, allowing the study of chemical or conformational exchange via NMR. An extension of this approach recently permitted the structural characterization of intermediates along the folding pathway of several proteins (Kay 2011b).

Another source of information regarding biomolecular dynamics comes from the measurement of amide-proton exchange rates, which provide data about global stability and local structural fluctuations. This chemical exchange with the solvent can be measured using diverse methods (Englander and Kallenbach 1983; Huyghues-Despointes et al. 2000). If the exchange is in a slow timescale (minutes to days), it can be determined by following the reduction in intensity of the NH signal in the HSQC of a protein dissolved in 100 % D_2O (the N-D peak is not observed in the spectrum). This experiment informs about the accessibility of each

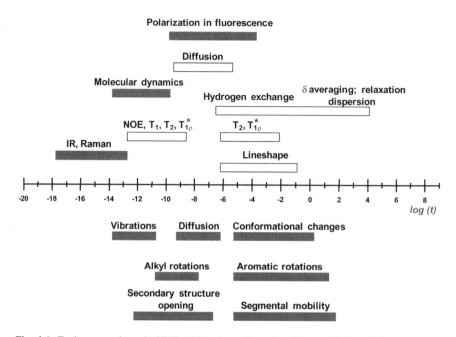

Fig. 4.6 Exchange regimes in NMR. White bars show the different NMR techniques used to map dynamics (the asterisk on the spin–lattice relaxation measurements indicates measurements of such relaxation time carried out within the rotating-frame, Chap. 3). In *grey* the motions which can be monitored by different biophysical techniques

backbone amide, correlating with structural information such as secondary and tertiary protein structure, and it provides data about the stability of particular hydrogen bonds. Furthermore, in specific conditions (high pH and temperature) it is possible to obtain information about the kinetic constant of the unfolding process (Sivaraman and Robertson 2001). If exchange takes place on a faster time-scale (ms), it can be observed by following the exchange of amide proton magnetization with the water protons, and a measurement of the dynamics of bound water molecules can be obtained (Otting et al. 1991).

4.5 Biomolecular Interactions (NMR in Drug Discovery)

Nuclear Magnetic Resonance is one of the main biophysical tools currently available for the study of drug-biomolecule interactions and, jointly with X-ray crystallography and EPR, able to provide detailed structural information. For this reason and because of the variety of applications that NMR offers it has recently become a powerful technique in the pharmaceutical industry. NMR presents advantages over other techniques used in the drug discovery process, among them its potential in the identification of very weak intermolecular interactions

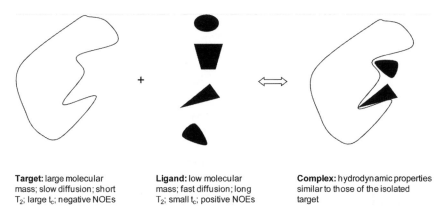

Target: large molecular mass; slow diffusion; short T_2; large t_c; negative NOEs

Ligand: low molecular mass; fast diffusion; long T_2; small t_c; positive NOEs

Complex: hydrodynamic properties similar to those of the isolated target

Fig. 4.7 Protein-ligand binding by NMR. Several ligands are represented against a generic target. The NMR properties for each of the molecules depending on their sizes are indicated. Upon binding, the NMR properties of the complex are similar to that of the isolated target

(mM range) that normally will pass undetected under the high-throughput screening (HTS) of in vitro enzyme bioassays. Another major points for utilising NMR as screening methodology is its ability to provide information at atomic resolution, of great value in later drug design and chemical evolution processes, as well as the versatility of experiments offered by the technique depending on the target/ligand requirements. Also in its favour, NMR does not require previous knowledge of the function of the protein, making it particularly useful in cases where no activity bio-assay is available. The NMR experiments commonly used in the drug-discovery process are based on the detection of those variations taking place on typical NMR parameters (chemical shifts, line-shapes, J-couplings and relaxation effects) upon the binding of target and ligand (Fig. 4.7). These differences are caused by the chemical environment changes affecting the involved molecules during the binding process. The interested reader can have a wider look on the different NMR techniques used to monitor molecular binding in references (Fielding 2007; Carlomagno 2005; Zerbe 2003).

4.5.1 Order of Affinities Measured

The *strength* of the target-ligand interaction is normally measured in terms of its dissociation constant, K_d, which is described as:

$$E + L \leftrightarrow EL\,;\; K_d = \frac{k_{on}}{k_{off}} = \frac{[E][L]}{[EL]}$$

meaning the tighter the binding, the lower the K_d.

Although NMR may not be the preferred biophysical technique for measuring binding affinities between biomolecules and small compounds, under appropriate circumstances it is capable of giving a good approximation of the dissociation constant K_d. NMR has the advantage of extending the measurable range of affinity constants into the *weak binding* region (K_d ranges above 10^{-5} M), a detection area that other screening techniques cannot cover. However, accurate measurement of dissociation constants requires concentrations of protein $<K_d$. Therefore NMR techniques cannot measure K_d values smaller than the limit of detection of the experiment (around 10–100 μM for routine NMR).

Interactions between target and ligand can be classed as slow, intermediate and fast exchange, with different methods to measure the K_d (Fig. 4.7). Systems in a regime of slow exchange will show two independent signals for the target, corresponding to its bound and free states. Slow exchange is due to slow rate of dissociation and therefore low K_d. The intensities of the signals from the protein in slow exchange will vary for different ligand concentrations (more ligand will increase the number of target molecules in the bound state). The K_d can then be measured by integrating the signals corresponding to both states at different known concentrations. If the system is in the intermediate exchange regime, the K_d can be evaluated by acquiring NMR spectra at different ligand concentrations. However the shape of the NMR signal, especially near or at the coalescence point (the point where the two signals converge into a single one), normally presents low resolution and reduced signal-to-noise, leading to estimates of the dissociation constant that are not fully reliable. The fast exchange regime ranges from 10 μM to >10 mM (weak binding), and shows the best attributes for a useful measurement of K_d via NMR. Changes in chemical shift or relaxation are normally the parameters used for such measurements, or any other factor that shows fast exchange behaviour.

4.5.2 Experiments in NMR Screening

The variety of methods that NMR spectroscopy offers for the study of the biomolecule-ligand interaction can be divided in two major groups depending on the molecule they focus on: target-based or ligand-based.

4.5.2.1 Target-Based Experiments

Target-based experiments examine the response of NMR parameters (most usually perturbations in the chemical shift values) of the biomolecule upon the mixing with the ligand(s). The best known method is the *SAR-by-NMR* technique that applies heteronuclear HSQC experiments to identify whether molecules bind to a protein (Hajduk et al. 1999). This technique necessitates the production of isotopically labelled macromolecules, ^{15}N-labelled if the amide groups of the protein are to be examined, or ^{13}C-labelled if the methyl groups of selected amino acids (e.g. leucine, valine, isoleucine) are monitored upon addition of potential binding ligands. Any molecule that binds to the biomolecule will alter the chemical environment of those residues responsible for the interaction, with the consequent modification of their chemical shift values in the HSQC. Although the ligand is added in excess to saturate the protein, isotope-filtered experiments (^{15}N or ^{13}C) allow detecting the biomolecule signals without interference from the ligand ones. One of the main advantages of the *SAR-by-NMR* over other biophysical techniques applied in drug discovery or even other NMR-based methods is that, apart from determining the presence or absence of binding, it renders very relevant site-specific structural information (i.e. the binding site of the protein) as each signal in the HSQC is directly associated to one particular amino acid in the protein sequence. Either because the HSQC has been previously assigned or by identifying the peaks corresponding to the binding-site region using a known binder to the target, it is possible to discern if a molecule binds to the same site or to a different one (Fig. 4.8a). This HSQC-based experiment traces protein resonances but not ligand ones, and therefore if a positive compound is detected within a cocktail of molecules, a subsequent investigation is required. The measurable affinity range is wider than for ligand-based methods (see below), especially at the high-affinity end. However, an important limitation of the *SAR-by-NMR* method comes from the target size, the best results being obtained for proteins below 40 kDa. Larger biomolecules increase the signal overlap and show less favourable relaxation properties leading to signal broadening, although the advent of the TROSY techniques has allowed an increase in the target size limit up to 70–80 kDa. Using the HSQC-based methods the K_d can be determined by titrating ligand into the target and measuring the changes in chemical shift as a function of target/ligand ratio.

4.5.2.2 Ligand-Based Experiments

The ligand-based NMR techniques applied to drug discovery monitor the changes in the NMR parameters of the ligand, e.g. chemical shift perturbations, relaxation parameters, diffusion coefficients, or inter/intramolecular magnetization transfers. In these methods, an excess of ligand over target is used, and the exchange between the bound and free states of the ligand originates the effects to be measured. In contrast to the above HSQC experiments, the biomolecule is required in much smaller quantities and no isotopic labelling is required. Additionally, the size of the protein is not an issue for this group of techniques, with some of the experiments clearly benefitting from large targets. Distinguishing specific from non-specific binding to the biomolecule is not as straightaway as in the HSQC, as no direct information from the binding site is obtained. Also, these techniques are unable to detect in a direct way high affinity ligands since the exchange between bound and free state for strong binders will not be fast enough during the NMR experiment to observe transfer of information (the ligand will spend too much time bound to the

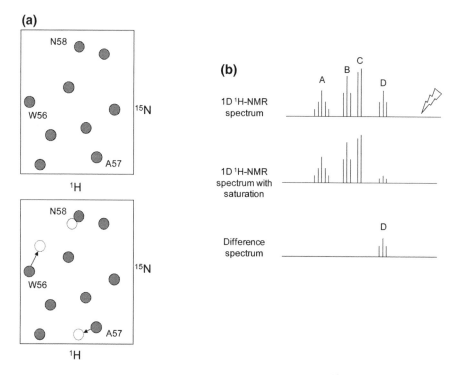

Fig. 4.8 NMR-based drug discovery. **a** *SAR-by-NMR* experiment: the ¹⁵N-HSQC spectrum of a protein in the absence of ligand is shown, with some of the signals labelled according to particular residues (*top*). Upon addition of the ligand some signals change their position due to the binding, while others remain unaltered. The *dotted circles* indicate the new positions, and the arrows show the shift of the changing signals. **b** STD-NMR: a schematic representation of the signals from four ligands (A–D) in the presence of the target is shown (the signals of the protein are not visible because the target is at low micromolar concentrations). The irradiation of the spectrum at a frequency close to protein signals but not to ligand ones (the bolt) leads to a spectrum where the intensity of some of the signals are decreased. The subtraction of the spectrum from a reference experiment irradiating at a frequency absent of near resonances provides those signals from the ligand that binds to the target (D)

target to transfer its effect to the free ligand). However, these experimental hurdles can be overcome by using competition experiments (see below).

Most of the ligand-based experiments register variations of ¹H NMR spectra, which are much faster to acquire than the 2D HSQCs used in the target-based techniques. One of the most commonly used experiments is the STD (Saturation Transfer Difference) (Mayer and Meyer 1999) (Fig. 4.8b). In this pulse sequence two different spectra are recorded: in the first one all the protons from the target are saturated via spin diffusion and the second is acquired without such saturation. Any ligand that binds to the biomolecule will be also saturated in the first experiment, transferring this saturation via exchange to the same ligand in free form. As a result, the difference between saturated/non-saturated spectra will yield a 1D ^{1}H-NMR showing just those signals corresponding to the bound ligands (the unbound molecules will yield the same spectrum either saturating or not the protein, and their resonances will cancel out) (Fig. 4.8b). The relative intensity of the signals from an interacting ligand provides information about the binding epitope, as the most intense resonances will arise from that part of the ligand in direct contact with the target due to a better magnetization transfer. The concentrations needed for the target molecule are relatively low (1–5 µM) and a large excess of ligand (approximately 100-fold) is normally used, as it allows more molecules of free ligand to be saturated and therefore a gain in sensitivity (and faster acquisition times). The detection limit for this method is around 10 mM, which is very advantageous in the search of weak binders. On the contrary, if the binding is tight (nM range) the saturation mechanism is not very efficient and ligands binding with a K_d in this range might result in false negatives.

Another ligand-based method that has found extensive application is the *water-LOGSY* experiment (water-Ligand Observation with Gradient SpectroscopY) which uses the water molecules bound at the target-ligand interface to transfer magnetization in a selective manner from the target-ligand complex to the free ligand (Dalvit et al. 2000). In the water-LOGSY, a positive interaction between ligand and target is characterized by positive ligand signals in the 1D ^{1}H-NMR spectrum or by a reduction in the negative signal obtained in the absence of the target (reference spectrum). Analogous to the STD experiment, the magnetization transfer is more efficient for large proteins with long τ_c. The target concentration used is also quite low (1–5 µM), with the best results obtained when the ligands are in a 20–1 excess ratio. The water-LOGSY is a very sensitive experiment for the screening of very weak binders, but like the STD, high affinity ligands cannot be detected with the standard experiment.

Other available methods are based on the nuclear Overhauser effect (NOE), like the *transferred NOE (trNOE)* experiment where the NOE signals of the ligand change sign if bound to the biomolecule. The trNOEs depend on the different tumbling times, τ_c, of the free and bound molecules. Small molecules have short correlation times and they show positive NOEs, whereas large molecules like proteins or nucleic acids exhibit strong negative NOEs (Chap. 2). If a small molecule binds to a protein, the former will tumble with the target when bound, showing strong negative NOEs or trNOEs. Therefore the sign and size of the NOEs can inform about the ligand–protein binding. Another NOE-based experiment is *NOE-pumping*, where magnetization transfer from the target to the ligand is observed via NOE. The methodology uses a "diffusion filter" which filters out the ligand magnetization while maintaining that of the target, taking advantage of the different sizes of the molecules involved. During the NOE mixing time the magnetization from the target will relax and part will be transferred to the ligand via intermolecular cross relaxation. Other ligand-based experiments observe variations in NMR parameters such as the effects on the relaxation of the small-molecule upon binding to the much larger biomolecule. These methods are also very sensitive, and afford similar information to the above ones, but their application in actual drug discovery projects has been much more limited.

In ligand-based techniques, the use of heteronuclei for the detection of binding between fragments and biomolecules has become increasingly popular. Its application is largely limited to the fluorine atom (^{19}F) due to its high sensitivity and 100 % natural abundance, although the ^{31}P nucleus has been shown to work in a similar way (Table 1.1). These experiments present some advantages over the typical ^{1}H-based ones: fewer NMR signals to analyse (usually one per ligand), a large chemical shift dispersion (reducing overlap) and high sensitivity to relaxation effects (^{19}F has a large CSA) which allows an easy detection of weak binders by observing the broadening of the heteronucleus signal (Dalvit 2007).

In general, one of the major drawbacks of the ligand-based methods is their inability to detect high affinity binders. To overcome this major obstacle several approaches have been described based on the introduction of a known ligand of the target (a reporter) that allows observing competition effects. If this known binder has a weak to medium affinity, these competition experiments will allow the detection of high affinity ligands, as any ligand binding on the same site with a similar or higher affinity for the target will affect somehow the NMR signal from the reporter. They also allow the identification of non-specific binders, as false positives should not cause any effect on the signals from the reporter as they do not share the same binding site at the target. The experiments most commonly used for competition binding are STD and water-LOGSY.

The use of the 1D ^{1}H NMR ligand-based methods allows the detection of binding using compound concentrations much below the K_d of the interaction, i.e. detecting binding affinity in the millimolar range while using concentrations in the micromolar range. Competition experiments can also be very useful for the quantitative measurement of affinities if the reporter has a known affinity constant (preferably in the weak to medium range), as this will allow the K_d ranking of new binding ligands. Another methodology to rank binding affinities using NMR monitors the line-broadening of the 1D ^{1}H spectra of the ligand (in excess) upon binding to the biomolecule. The sharp NMR signals of a small molecule are broadened by binding to a high molecular weight macromolecule, with a subsequent decrease in signal intensity.

4.6 Other Applications

The intrinsic flexibility of the magnetic resonance technique has allowed its expansion into many different scientific fields such as molecule characterization, biochemistry, imaging or drug discovery. In this sense, we can compare NMR to a toolbox that is constantly being added to with brand new *tools* specifically designed for new *problems* that need fixing. Some experiments like the 1D ^{1}H or the 2D COSY or HSQC will work as our screwdrivers and hammer, tools that are constantly needed and that we normally have to hand. Other experiments like the 3D HNCO are problem-specific, and will be used only in restricted fields like biomolecular NMR. And if we do not have the tool required to fix a particular problem, we shall be able to fabricate a new one from scratch or modify an existing tool to our requirements. As a case in point of the utilities in the NMR toolbox, we shall look in this section at two that have become fields of their own: metabolomics and solid-state NMR.

4.6.1 Metabolomics

The term *metabolomics* or *metabonomics* defines the comprehensive and simultaneous study of the small molecules that participate in metabolism and the changes they experience in response to different biological actions. Understanding how metabolic processes proceed sheds light on pathologies, clinical diagnosis, toxicology, genomics, biomarker identification, etc. Alongside mass spectrometry, NMR spectroscopy has been extensively used as analytical tool in metabolic studies (Lindon et al. 2007). Foremost among the benefits that the use of NMR has brought to metabolomics is its applicability to all types of samples, from biological tissues to fluids like blood, urine or saliva, plant or cell extracts, providing insight into the condition of living organisms. Metabolomics deals with very complex mixtures, and NMR has proven itself a very powerful methodology for the identification and structural characterization of new metabolites. NMR has the advantage of being a non-invasive and non-destructive technique; therefore the sample can be recovered at the end of the experiment. Sample preparation is relatively easy and due to the normally large number of samples to measure and experiment repetition, automation systems are normally used.

In general, the experimental procedure of NMR metabolomics is based on the acquisition of 1D ^{1}H spectra of the available samples. In order to avoid any interference that could affect the analysis, it is extremely important to apply an established protocol to all the stages previous to the NMR step: sample collection, storage or pre-NMR manipulation. Also, the NMR part of the metabolomics study must be carried out under identical experimental conditions for all samples: temperature calibration, pulse/power parameters, number of scans, data processing or any other step in the acquisition and processing of the raw data, in order to eliminate any factors that could bias the analysis afterwards. Although, in theory, a typical 1D ^{1}H spectrum would render the profile of metabolite signals needed for the metabolomics study, two variants are normally favoured: (i) the 1D version of the nuclear Overhauser enhancement spectroscopy (NOESY) with presaturation of the water signal and (ii) the T_2 relaxation spin-echo 1D (Carr-Purcell-Meiboom-Gill) CPMG sequence (similar to the echo sequence used to measure the T_2 relaxation time, Fig. 1.6b). The 1D NOESY provides a full ^{1}H spectrum with no signal attenuation and good water suppression, while the relaxation delay of the CPMG is optimized for the suppression of the broad signals of the spectrum arising from large molecules such as proteins and lipids, and in this way simplifying the metabolite profile. Another experiment that has found some use is the diffusion-edited 1D ^{1}H which is based on the same principle as the CPMG, but in this case it is optimised to eliminate signals from the smaller metabolites from the spectrum while keeping the ones arising from large molecules.

For complete metabolite identification and structural characterization some additional experiments are required, although due to their longer duration they are normally applied only to selected samples. Thus, conventional 2D NMR methods such as homonuclear COSY and TOCSY, as well as ^{1}H-^{13}C correlation experiments such as HMQC, HSQC and HMBC (Chap. 3) are commonly acquired in metabolomics studies. Another widespread NMR technique in metabolomics is the 2D *J*-resolved pulse sequence, where the coupling patterns for each of the signals present in the 1D ^{1}H splits in a two dimensional spectrum. The identification of metabolites is carried out combining all the above 1D and 2D experiments, as well as the information collected in metabolite NMR databases such as the Biological Magnetic Resonance Data Bank (BMRB, http://www.bmrb.wisc.edu) or the Human Metabolome Database (HMDB, http://www.hmdb.ca).

The comparison of the metabolic profile between the different samples is commonly done by dividing the chemical shift of the 1D ^{1}H spectra (NOESY or CPMG) into small "bins" or "buckets" of normally 0.01–0.04 ppm width. This approach minimizes the problem of small chemical shift differences that could arise within a group of samples due to variable conditions such as pH or electrolyte content. The total peak area or signal intensity within these buckets is added up and compared among the samples using different statistical analysis tools. One of the main statistics method applied in metabolomics is Principal Component Analysis or PCA, an unsupervised statistical technique that allows the reduction of an NMR spectra comprising many variables (the bins) into a lower dimensional PCA space. In this way, the entire 1D ^{1}H spectrum is converted into a single point in a PCA scores plot, where the clustering of points resembles similarity between NMR spectra/samples (i.e. similar metabolic profile), while the separation of points will indicate one or several metabolite differences.

4.6.2 *Solid State NMR and HR-MAS*

Most of this book has been dedicated to the NMR basics and the description of essential applications of the technique in the solution state, which is where NMR has found the widest range of uses. However, not every material can be studied in solution either because they are unstable, insoluble or because the properties we are interested in can only be analysed in the solid state. Although not as routinely utilised as solution NMR, solid-state NMR (ssNMR) has found wide applicability, especially in the study of inorganic materials (e.g. zeolites, polymers) and also in the field of biological systems such as tissues, membrane proteins, fibres and protein aggregates. ssNMR is able to provide relevant information regarding structure, morphology, heterogeneity or dynamic processes of these samples. In the case of ssNMR, an increase in the molecular weight of the molecule does not affect the line-shape of the signals, contrary to what we have seen in solution NMR.

The molecular Brownian motions in solution have the effect of averaging to zero several intrinsic properties of the nuclei, thus cancelling out any possible effects in the NMR spectra (Chap. 1). In the solid-state the orientation of molecules is not random but ordered, and different parameters have to be taken into account for a full comprehension of the NMR data. For instance, the NMR spectra of any molecule acquired both in solution and in the solid-state will show apparent differences in terms of signal resolution. The most conspicuous is the very broad signals observed in the solid state occupying the whole chemical shift range of the nuclei and without a clear baseline. This broadening is caused by anisotropic spin interactions that are not averaged to zero in the solid state. There are different physical mechanisms that contribute to the broadening, the dominant one being the dipole–dipole interaction affecting the local field seen by the nuclei, as expressed by the following equation for a generic spin I:

$$B_{loc} = \pm \mu_S r_{IS} - 3\left(3\cos^2\theta_{IS} - 1\right)$$

where μ is the magnetic moment of nuclei S, r the internuclear distance between the I and S, and θ the angle between the internuclear vector and the local field (Fig. 1.3). The angular dependence is not averaged in solids and leads to the broad signals commonly observed. Also, depending on the values of the magnitude described above, the splitting caused by the dipole–dipole interaction can reach figures in the range of tens of kilohertzs. Relaxation via the chemical shift anisotropy (CSA) mechanism also contributes to broadening, although to a lesser extent, which also presents a variation with the θ angle with the same $3\cos^2\theta_{IS}$-1 dependence.

One of the main implementations to eliminate the angle dependence in solids NMR is via the application of the *magic angle spinning* or MAS. The NMR sample rotates at an angle of 54° 44′ (Fig. 4.9a) resulting in the factor $3\cos^2\theta$-1 taking a zero value, thus *magically* removing the angle parameter from both the dipole–dipole and CSA broadening mechanisms and producing much sharper NMR signals. In practical terms, for the MAS to produce good results the rotation frequency must be of the order of the CSA linewidth (i.e. a few kilohertzs, 2–20 kHz). Dipolar broadening also benefits from the MAS application, although it is commonly necessary to include a decoupling block at the end of the pulse sequences to achieve complete elimination. Because of the large dipolar interaction the power needed for the decoupling is much higher than the values used normally in solution-state NMR and different methods are used to avoid overheating the sample.

Solid-state NMR suffers from the same insensitivity as solution NMR experiments, and several approaches are applied in order to enhance the NMR signal in solids. One of the most popular is the *cross polarization* (CP) method, which achieves the transfer of polarization from a high γ nucleus to a low γ one using the dipolar coupling between them (Fig. 4.9b). The source of polarization is typically the very sensitive ^1H spin, and by using the CP the sensitivity of the nuclei S to which it is dipolarly coupled can be enhanced by a factor of up to γ_H/γ_S. The CP pulses extend for several milliseconds in contrast to the microsecond duration used in solution state experiments. During these CP pulses, cross relaxation is transferred between the coupled nuclei such as ^1H and ^{13}C or ^{15}N, with the effect of increasing the magnetization of the heteronuclei. Because the T_1 of the ^1H is generally much shorter than that of the S nucleus, the pulse sequence can be repeated much faster (more accumulations) with the consequent improvement in sensitivity. For the CP to work experimentally, a match has to be reached between the radiofrequency fields of the ^1H and the S nucleus (the called Hartmann-Hahn condition): $\gamma_H B_H = \gamma_S B_S$. Additional sensitivity enhancements can be obtained by including dipolar decoupling on the ^1H frequency channel during FID recording as well as applying MAS to sharpen the NMR signals.

Many solid-state NMR methods seek to eliminate the original causes of the broad signals, such as the dipole–dipole and CSA couplings. However, those

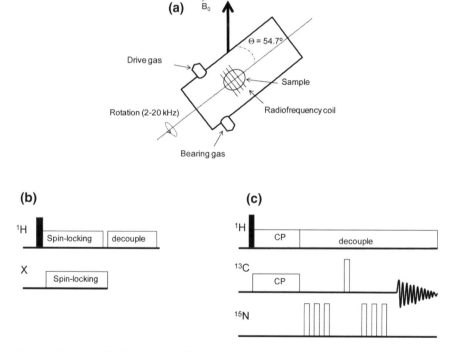

Fig. 4.9 Solid-state NMR. **a** Scheme of the measurements in MAS (magic angle spinning). **b** Cross polarization: the effect of the sequence is to transfer magnetization from the abundant ^1H spins in the sample to the X spin via the dipolar coupling between ^1H and X spins. The spin-lock fields are also called "contact pulses". **c** REDOR sequence: the equally spaced 180° ^{15}N pulses result in dephasing of transverse carbon magnetization produced by the CP transfer from dipolar coupled protons. The phase of the pulses and the spin-lock fields are described elsewhere (Duer 2004)

couplings contain significant structural information that we might be interested to retain. If this is the case, a dipolar *recoupling* can be applied in experiments. To achieve recoupling we synchronize the pulse sequence with the MAS rotation speed, reintroducing selectively the dipolar coupling while, at the same time, maintaining the sharp signals generated by the MAS. The intensity of the NMR signals with recoupling correlates with the duration of the recoupling sequence and also with the distance of those nuclei dipolarly coupled. Therefore, using this method internuclear distances can be determined, as well as bond and torsional angles. The most widespread sequence for dipolar recoupling is the rotational-echo double resonance or REDOR (Fig. 4.9c) where several 180° pulses are synchronized with the rotation of the sample to reintroduce the heteronuclear coupling, allowing the acquisition of experiments similar to those obtained for liquids such as HMQC, HSQC or NOESY.

It can be said that ssNMR has come of age and it offers excellent and fascinating applications for example in the field of membrane proteins, or even in tackling the structure of whole intact virus (Park et al. 2010; Goldbourt et al. 2010). The interested reader in a more profound description of ssNMR can have a good introduction in the literature (Duer 2004).

References

Allen M, Varani L, Varani G (2001) Nuclear Magnetic Resonance methods to study structure and dynamics of RNA-protein complexes. Methods Enzymol 339:357–376

Arrondo JLR, Goñi FM (1999) Structure and dynamics of membrane proteins as studied by infrared spectroscopy. Prog Biophys Mol Biol 72:367–405

Carlomagno T (2005) Ligand-target interactions: what can we learn from NMR? Ann Rev Biophys Biomol Struct 34:245–266

Cavanagh J, Fairbrother WJ, Palmer AG III, Skelton NJ (1996) Protein NMR spectroscopy: theory and practice. Academic Press, New York

Dalvit C, Pevarello P, Tato M, Veronesi M, Vulpetti A, Sundstrom M (2000) Identification of compounds with binding affinity to proteins via magnetization transfer from bulk water. J Biomol NMR 18:65–68

Dalvit C (2007) Ligand- and substrate-based F-19 NMR screening: principles and applications to drug discovery. Prog Nucl Magn Reson Spectrosc 51:243–271

Duer MJ (2004) Introduction to solid state NMR spectroscopy. Blackwell, London

Englander SW, Kallenbach NR (1983) Hydrogen exchange and structural dynamics of proteins and nucleic acids. Q Rev Biophys 16:521–655

Esfandiary R, Hunjan JS, Lushington GH, Joshi SB, Middaugh CR (2009) Temperature dependent 2nd derivative absorbance spectroscopy of aromatic amino acids as a probe of protein dynamics. Protein Sci 18:2603–2614

Fielding L (2007) NMR methods for the determination of protein-ligand dissociation constants. Prog Nucl Magn Reson Spectrosc 51:219–242

Gardner KH, Kay LE (1998) The use of ^{2}H, ^{13}C and ^{15}N multidimensional NMR to study the structure and dynamics of proteins. Annual Rev Biophys Biomol Struct 27:357–406

Goldbourt A, Day LA, McDermott AE (2010) Intersubunit hydrophobic interactions in Pf1 filamentous phage. J Biol Chem 285:37051–37059

Graslund S, Nördlund P et al (2008) Protein production and purification. Nat Methods 5:135–146

Grzesiek S, Bax A (1992) Improved 3D triple-resonance NMR techniques applied to a 31 kDa protein. J Magn Reson 96:432–440

Hajduk PJ, Meadows RP, Fesik SW (1999) NMR-based screening in drug discovery. Q Rev Biophys 32:211–240

Huyghues-Despointes BMP, Pace CN, Englander SW, Scholtz JM (2000) Measuring the conformational stability of a protein by hydrogen exchange. In: Murphy KE (ed) Protein structure, stability and folding. Methods in molecular biology, Vol. 168. Humana Press, Totowa

Ikura M, Kay LE, Bax A (1991) Improved three-dimensional 1H–13C-1H correlation spectroscopy of a 13C-labeled protein using constant-time evolution. J Biomol NMR 1:299–304

Kay LE, Ikura M, Tschudin R, Bax A (1990) Three-dimensional triple-resonance spectroscopy of isotopically enriched proteins. J Magn Reson 89:496–514

Kay LE (2011a) Solution NMR spectroscopy of supramolecular systems, why bother? A methyl-TROSY NMR view. J Magn Reson 210:159–170

Kay LE (2011b) NMR studies of protein structure and dynamics. J Magn Reson 213:492–494

Kime MJ (1984) Assignment of resonances in the *Escherichia coli* 5 S RNA fragment proton NMR spectrum using uniform nitrogen-15 enrichment. FEBS Lett 173:342–346

Lakowicz JR (2010) Principles of fluorescence spectroscopy, 3rd edn., corrected Springer, New York

Lindon JC, Nicholson JK and Holmes E. Elsevier B.V (eds) (2007) The Handbook of Metabonomics, Netherlands

Lipari G, Szabo A (1982a) Model-free approach to the interpretation of nuclear magnetic resonance relaxation in macromolecules. 1. Theory and range of validity. J Am Chem Soc 104:4559–4570

Lipari G, Szabo A (1982b) Model-free approach to the interpretation of nuclear magnetic resonance relaxation in macromolecules. 2. Analysis of experimental results. J Am Chem Soc 104:4546–4559

Lipsitz RS, Tjandra N (2004) Residual dipolar couplings in NMR structure analysis. Annual Rev Biophys Biomol Struct 33:387–413

Lundström P, Vallurupalli P, Hansen DF, Kay LE (2009) Isotopic labelling methods for the studies of excited protein states by relaxation dispersion NMR spectroscopy. Nat Protocols 4:1641–1648

Mayer M, Meyer B (1999) Characterization of ligand binding by saturation transfer difference NMR spectroscopy. Angew Chem Int Ed Engl 38:1784–1788

Mossakowska DE, Smith RAG (1997) Production and characterization of recombinant proteins for NMR structural studies. In: Reid DG (ed) Protein NMR techniques methods in molecular biology, Vol 60. Humana Press, Totowa

Muhandiram DR, Kay LE (1994) Gradient-enhanced triple-resonance three-dimensional NMR experiments with improved sensitivity. J Magn Reson 103:203–216

Neuhaus D, Williamson MP (1999) The nuclear Overhauser effect in structural and conformational analysis (2nd Ed). VCH Publishers, New York

Olejniczak ET, Xu RX, Fesik SW (1992) A 4D HCCH-TOCSY experiment for assigning the side chain 1H and 13C resonances of proteins. J Biomol NMR 2:655–659

Olsen JI, Schweizer MP et al (1982) Carbon-13 NMR relaxation studies of pre-melt structural dynamics in [4-13C-uracil] labeled E. coli transfer RNA1 Val* Nucl Acid Res 10:4449–4464

Otting G, Liepinsh E, Wüthrich K (1991) Protein hydration in aqueous solution. Science 254:974–980

Park SH, Marassi FM, Black D, Opella SJ (2010) Structure and dynamics of the membrane-bound form of Pf1 coat protein: implications of structural rearrangements for virus assembly. Biophys J 99:1465–1474

Patel DJ, Shapiro L, Hare D (1987) DNA and RNA: NMR studies of conformations and dynamics. Q Rev Biophys 20:35–112

Petsko GA, Ringe D (1984) Fluctuations in protein structure from X-ray diffraction. Ann Rev
    Biophys Bioeng 13:331–371
Price WS (1999) Water signal suppression in NMR spectroscopy. Annual Rep NMR Spectrosc
    38:289–354
Religa TL, Kay LE (2010) Optimal methyl labeling for studies of supra-macromolecular systems.
    J Biomol NMR 47:163–169
Riek R, Pervushin K, Wüthrich K (2000) TROSY and CRINEPT: NMR with large molecular and
    supramolecular structures in solution. Trends Biochem Sci 25:462–468
Ringe D, Petsko GA (1985) Mapping protein dynamics by X-ray diffraction. Prog Biophys Mol
    Biol 45:197–235
Ruschak AM, Velyvis A, Kay LE (2010) A simple strategy for ^{13}C, ^{1}H labeling at the Ile-γ2
    methyl position in highly deuterated proteins. J Biomol NMR 48:129–135
Sivaraman T, Robertson AD (2001) Kinetics of conformational fluctuations by EX1 hydrogen
    exchange in native proteins. Methods Mol Biol 168:193–214
Tinoco I Jr, Sauer K, Wang JC (2002) Physical chemistry: principles and applications in
    biological sciences, 4th edn. Prentice Hall, New Jersey
Tugarinov V, Hwang PM, Kay LE (2004) Nuclear magnetic resonance spectroscopy of high-
    molecular weight complexes. Ann Rev Biochem 73:107–146
Wüthrich K (1986) NMR of proteins and nucleic acids. Wiley and Sons, New York
Xu J, Plaxco KW, Allen SJ (2006) Probing the collective vibrational dynamics of a protein in
    liquid water by terahertz absorption spectroscopy. Protein Sci 15:1175–1181
Zerbe O (ed) (2003) BioNMR in Drug Research. Wiley-VCH, Weinheim

44980465R10073

Made in the USA
Middletown, DE
21 June 2017